PROBLEMS OF AGRICULTURAL GROWTH IN INDIA

PROBLEMS OF AGRICULTURAL GROWTH IN INDIA

Manish Dubey

CYBER TECH PUBLICATIONS
4264/3, Ansari Road, Daryaganj, New Delhi-110002 (India)
Ph.: 011-23244078, 011-43559448 Fax: 011-23280028
E-Mail: cyberpublicationsdelhi@yahoo.com
Website: www.cybertechpublications.com

PROBLEMS OF AGRICULTURAL GROWTH IN INDIA

Manish Dubey

First Edition 2011

Published by :

G.S. Rawat for **Cyber Tech Publications**
4264/3, Ansari Road, Daryaganj, New Delhi-110002 (India)
Ph.: 011-23244078, 011-43559448 Fax: 011-23280028
E-Mail: cyberpublicationsdelhi@yahoo.com
Website: www.cybertechpublications.com

CONTENTS

Preface

Several wild cereals, including rice, grew in the Vindhyan Hills, and rice cultivation, at sites such as Chopani-Mando and Mahagara, may have been underway as early as 7000 BP. The relative isolation of this area and the early development of rice farming imply that it was developed indigenously. Chopani-Mando and Mahagara are located on the upper reaches of the Ganges drainage system and it is likely that migrants from this area spread rice farming down the Ganges valley into the fertile plains of Bengal, and beyond into south-east Asia.

Irrigation was developed in the Indus Valley Civilization by around 4500 BCE. The size and prosperity of the Indus civilization grew as a result of this innovation, which eventually led to more planned settlements making use of drainage and sewers. Sophisticated irrigation and water storage systems were developed by the Indus Valley Civilization, including artificial reservoirs at Girnar dated to 3000 BCE, and an early canal irrigation system from circa 2600 BCE.

A variety of tropical fruit such as mango and muskmelon are native to the Indian subcontinent. The Indians also domesticated hemp, which they used for a number of applications including making narcotics, fiber, and oil. The farmers of the Indus Valley grew peas, sesame, and dates. Sugarcane was originally from tropical South Asia and Southeast Asia. Different species likely originated in different locations with *S. barberi* originating in India and *S. edule* and *S. officinarum* coming from New Guinea. This period also saw the first domestication of the elephant. Barley and wheat cultivation—along with the domestication of cattle, primarily sheep and goat—was visible in Mehrgarh by 8000-6000 BCE. Agro pastoralism in India included threshing, planting crops in rows—either of two or of six—and storing grain in granaries. By the 5th millennium BCE agricultural communities became widespread in Kashmir. Zaheer Baber (1996) states that 'the first evidence of cultivation of cotton had already developed'. Cotton was cultivated by the 5th millennium BCE-4th millennium BCE. The Indus cotton industry was well developed and some methods used in cotton spinning and fabrication continued to be practiced till the modern Industrialization of India.

— *Author*

1

History of Agriculture in India

Indian agriculture began by 9000 BCE as a result of early cultivation of plants, and domestication of crops and animals. Settled life soon followed with implements and techniques being developed for agriculture. Double monsoons led to two harvests being reaped in one year. Indian products soon reached the world via existing trading networks and foreign crops were introduced to India. Plants and animals—considered essential to their survival by the Indians—came to be worshiped and venerated.

The middle ages saw irrigation channels reach a new level of sophistication in India and Indian crops affecting the economies of other regions of the world under Islamic patronage. Land and water management systems were developed with an aim of providing uniform growth. Despite some stagnation during the later modern era the independent Republic of India was able to develop a comprehensive agricultural program.

Early History

Wheat, barley and jujube were domesticated in the Indian subcontinent by 9000 BCE; Domestication of sheep and goat soon followed. This period also saw the first domestication of the elephant. Barley and wheat cultivation—along with the domestication of cattle, primarily sheep and goat—was visible in Mehrgarh by 8000-6000 BCE. Agro pastoralism in India included threshing, planting crops in rows—either of two or of six—and storing grain in granaries. By the 5th millennium BCE agricultural communities became widespread in Kashmir. Zaheer Baber (1996) states that 'the first evidence of cultivation of cotton had already developed'. Cotton was cultivated by the 5th millennium BCE-4th millennium BCE. The Indus cotton industry was well developed and some methods used in cotton spinning and fabrication continued to be practiced till the modern Industrialization of India.

A variety of tropical fruit such as mango and muskmelon are native to the Indian subcontinent. The Indians also domesticated hemp, which they used for a number of applications including making narcotics, fiber, and oil. The farmers of the Indus Valley grew peas, sesame, and dates. Sugarcane was originally from tropical South Asia and Southeast Asia. Different species likely originated in different locations with *S. barberi* originating in India and *S. edule* and *S. officinarum* coming from New Guinea.

Wild Oryza rice appeared in the Belan and Ganges valley regions of northern India as early as 4530 BCE and 5440 BCE respectively. Rice was cultivated in the Indus Valley Civilization. Agricultural activity during the second millennium BC included rice cultivation in the Kashmir and Harrappan regions. Mixed farming was the basis of the Indus valley economy. Denis J. Murphy (2007) details the spread of cultivated rice from India into South-east Asia:

Several wild cereals, including rice, grew in the Vindhyan Hills, and rice cultivation, at sites such as Chopani-Mando and Mahagara, may have been underway as early as 7000 BP. The relative isolation of this area and the early development of rice farming imply that it was developed indigenously. Chopani-Mando and Mahagara are located on the upper reaches of the Ganges drainage system and it is likely that migrants from this area spread rice farming down the Ganges valley into the fertile plains of Bengal, and beyond into south-east Asia.

Irrigation was developed in the Indus Valley Civilization by around 4500 BCE. The size and prosperity of the Indus civilization grew as a result of this innovation, which eventually led to more planned settlements making use of drainage and sewers. Sophisticated irrigation and water storage systems were developed by the Indus Valley Civilization, including artificial reservoirs at Girnar dated to 3000 BCE, and an early canal irrigation system from circa 2600 BCE.

Archeological evidence of an animal-drawn plough dates back to 2500 BC in the Indus Valley Civilization.

VEDIC PERIOD – POST MAHA JANAPADAS PERIOD (1500 BCE – 200 CE)

Jute was first cultivated in India, where it was used to make ropes and cordage. Some animals—thought by the Indians as being vital to their survival—came to be worshiped. Trees were also domesticated, worshiped, and venerated—*Pipal* and *Banyan* in particular. Others came to be known for their medicinal uses and found mention in the holistic medical system *Ayurveda.*

Gupta (2004) finds it likely that summer monsoons may have been longer and may have contained moisture in excess than required for normal food production. One effect of this excessive moisture would have been to aid the winter monsoon rainfall required for winter crops. In India, both wheat and barley are held to be *Rabi* (winter) crops and—like other parts of the world—would have largely depended on winter monsoons before the irrigation became widespread. The growth of the *Kharif* crops would have probably suffered as a result of excessive moisture.

The Encyclopedia Britannica—on the subject of agriculture of the later Vedic period—holds that:

In the later Vedic texts (c. 1000–500 BC), there are repeated references to iron. Cultivation of a wide range of cereals, vegetables, and fruits is described. Meat and milk products were part of the diet; animal husbandry was important. The soil was plowed several times. Seeds were broadcast. Fallowing and a certain sequence of cropping were recommended. Cow dung provided the manure. Irrigation was practiced.

The Mauryan Empire (322–185 BCE) categorized soils and made meteorological observations for agricultural use. Other Mauryan facilitation included construction and maintenance of dams, and provision of horse-drawn chariots—quicker than traditional bullock carts.

The Greek diplomat Megasthenes (c. 300 BC)—in his book *Indika*— provides a secular eyewitness account of Indian agriculture:

India has many huge mountains which abound in fruit-trees of every kind, and many vast plains of great fertility.... The greater part of the soil, moreover, is under irrigation, and consequently bears two crops in the course of the year.... In addition to cereals, there grows throughout India much millet... and much pulse of different sorts, and rice also, and what is called bosporum [Indian millet]. Since there is a double rainfall [i.e., the two monsoons] in the course of each year... the inhabitants of India almost always gather in two harvests annually.

EARLY COMMON ERA – HIGH MIDDLE AGES (200–1200 CE)

Spice trade involving spices native to India—including cinnamon and black pepper—gained momentum as India starts shipping spices to the Mediterranean. Roman trade with India followed as detailed by the archaeological record and the *Periplus of the Erythraean Sea*. Chinese sericulture attracted Indian sailors during the early centuries of the common era.

The Tamil people cultivated a wide range of crops such as rice, sugarcane, millets, black pepper, various grams, coconuts, beans, cotton, plantain, tamarind and sandalwood. Jackfruit, coconut, palm, areca and plantain trees were also known. Systematic ploughing, manuring, weeding, irrigation and crop protection was practiced for sustained agriculture. Water storage systems—some of the earliest in the world—were designed during this period. Kallanai (1st-2nd Century CE), a dam built on river Kaveri during this period, is considered the oldest water-regulation structure in the world still in use.

Crystallized sugar was discovered by the time of the Imperial Guptas (320-550 CE), and the earliest reference of candied sugar come from India. The process was soon transmitted to China with traveling Buddhist monks. Chinese documents confirm at least two missions to India, initiated in 647 CE, for obtaining technology for sugar-refining. Each mission returned with results on refining sugar. Indian spice exports find mention in the works of Ibn Khurdadhbeh (850), al-Ghafiqi (1150), Ishak bin Imaran (907) and Al Kalkashandi (fourteenth century).

Noboru Karashima's research of the agrarian society in South India during the Chola Empire (875-1279) reveals that during the

Chola rule land was transferred and collective holding of land by a group of people slowly gave way to individual plots of land, each with their own irrigation system. The growth of individual disposition of farming property may have led to a decrease in areas of dry cultivation. The Cholas also had bureaucrats which oversaw the distribution of water—-particularly the distribution of water by tank-and-channel networks to the drier areas.

LATE MIDDLE AGES – EARLY MODERN ERA (1200–1757 CE)

The construction of water works and aspects of water technology in India is described in Arabic and Persian works. The diffusion of Indian and Persian irrigation technologies gave rise to an advanced irrigation system which bought about economic growth and growth of material culture.

Agricultural 'zones' were broadly divided into those producing rice, wheat or millets. Rice production continued to dominate Gujarat and wheat dominated north and central India.

The Encyclopedia Britannica further details the many crops introduced to India during this period of extensive global discourse:

Introduced by the Portuguese, cultivation of tobacco spread rapidly. The Malabâr Coast was the home of spices, especially black pepper, that had stimulated the first European adventures in the East. Coffee had been imported from Abyssinia and became a popular beverage in aristocratic circles by the end of the century. Tea, which was to become the common man's drink and a major export, was yet undiscovered, though it was growing wild in the hills of Assam. Vegetables were cultivated mainly in the vicinity of towns. New species of fruit, such as the pineapple, papaya, and cashew nut, also were introduced by the Portuguese. The quality of mango and citrus fruits was greatly improved.

Land management was particularly strong during the regime of Akbar the Great (reign: 1556-1605), under whom scholar-bureaucrat Todarmal formulated and implemented elaborated methods for agricultural management on a rational basis. Indian crops—such as cotton, sugar, and citric fruits—spread visibly throughout North Africa, Islamic Spain, and the Middle East. Though they may have been in cultivation prior to the solidification of Islam in India, their production was further improved as a result of this recent wave, which led to far-reaching economic outcomes for the regions involved.

COLONIAL BRITISH ERA (1757–1947 CE)

Few Indian commercial crops made it to the global market under the British Raj in In India. Cotton, indigo, opium, and rice were known in particular. The second half of the 19th century saw some increase in land under cultivation, and agricultural production expanded at an average rate of about 1 percent per year by the later 19th century. Due to extensive irrigation by canal networks Punjab, Narmada valley, and Andhra Pradesh became centers of agrarian reforms.

Roy (2006) comments on the Influence of the world wars on the Indian agricultural system: Agricultural performance in the interwar period (1918–1939) was dismal. From 1891 to 1946, the annual growth rate of all crop output was 0.4 percent, and food-grain output was practically stagnant. There were significant regional and intercrop differences, however, nonfood crops doing better than food crops. Among food crops, by far the most important source of stagnation was rice. Bengal had below-average growth rates in both food and nonfood crop output, whereas Punjab and Madras were the least stagnant regions. In the interwar period, population growth accelerated while food output decelerated, leading to declining availability of food per head. The crisis was most acute in Bengal, where food output declined at an annual rate of about 0.7 percent from 1921 to 1946, when population grew at an annual rate of about 1 percent.

The British regime in India did supply the irrigation works but rarely on the scale required. Community effort and private investment soared as market for irrigation developed. Agricultural prices of some commodities rose to about three times between 1870-1920.

REPUBLIC OF INDIA (1947 CE ONWARDS)

Special programs were undertaken to improve food and cash crops supply. The Grow More Food Campaign (1940s) and the Integrated Production Programme (1950s) focused on food and cash crops supply respectively. Five-year plans of India—oriented towards agricultural development—soon followed. Land reclamation, land development, mechanization, electrification, use of chemicals—fertilizers in particular, and development of agriculture oriented 'package approach' of taking a set of actions instead of promoting single aspect soon followed under government supervision. The many 'production revolutions' initiated from 1960s onwards included Green Revolution in India, Yellow Revolution (oilseed: 1986-1990), Operation Flood (dairy: 1970-1996), and Blue Revolution (fishing: 1973-2002) etc. Following the economic reforms of 1991, significant growth was registered in the agricultural sector, which was by now benefiting from the earlier reforms and the newer innovations of Agro-processing and Biotechnology.

Due to the growth and prosperity that followed India's economic reforms a strong middle class has emerged as the main consumer of fruits, dairy, fish, meat and vegetables—a marked shift from the earlier staple based consumption. Since 1991, changing consumption patterns have led to a 'revolution' in 'high value' agriculture while the need for cereals is experiencing a decline. In fact, the per capita consumption of cereals declined from 192 to 152 kilograms from 1977 to 1999 while the consumption of fruits increased by 553%, vegetables by 167%, dairy products by 105%, and non-vegetarian products by 85% in India's rural areas alone. Urban areas experienced a similar increase.

Agricultural exports continued to grow at well over 10.1% annually through the 1990s. Contract farming—which requires the farmers to produce crops for a company under contract—and high value agricultural product increased. Contract farming led to a decrease in transaction costs while the contract farmers made more profit compared to the non-contract workforce. However, small landholding continued to create problems for India's farmers as the limited land resulted in limited produce and limited profits.

Since independence, India has become one of the largest producers of wheat, edible oil, potato, spices, rubber, tea, fishing, fruits, and vegetables in the world. The Ministry of Agriculture oversees activities relating to agriculture in India. Various institutions for agriculture related research in India were organized under the Indian Council of Agricultural Research (est. 1929).

Other organizations such as the National Dairy Development Board (est. 1965), and National Bank for Agriculture and Rural Development (est. 1982) aided the formation of cooperatives and improved financing. The contribution of agriculture in employing India's male workforce declined from 75.9% in 1961 to 60% in 1999–2000. Dev (2006) holds that 'there were about 45 million agricultural labor households in the country in 1999–2000.' These households recorded the highest incidence of poverty in India from 1993 to 2000.

During 2003-04, agriculture accounted for 22 % of India's GDP and employed 58 per cent of the country's workforce. India is the world's largest producer of milk, fruits, cashew nuts, coconuts, ginger, turmeric, banana, sapota, pulses, and black pepper. India is the second largest producer of groundnut, wheat, vegetables, sugar and fish in the world. India is also the third largest producer of tobacco and rice, the fourth largest producer of coarse grains, the fifth largest producer of eggs, and the seventh largest producer of meat. The green revolution introduced high yielding varieties of crops which also increased the usage of fertilizers and pesticides. About 90% of the pesticide usage in India is accounted for by DDT and Lindane (BHC/HCH). There has been a shift to organic agriculture particularly for exported commodities.

GREEN REVOLUTION IN INDIA

The introduction of high-yielding varieties of seeds after 1965 and the increased use of fertilizers and irrigation are known collectively as the Green Revolution, which provided the increase in production needed to make India self-sufficient in food grains, thus improving agriculture in India. Famine in India, once accepted as inevitable, has not returned since the introduction of Green Revolution crops.

Initiation

The program was started with the help of the United States-based Rockefeller Foundation and was based on high-yielding varieties of wheat, rice, and other grains that had been developed in Mexico and in the Philippines. Of the high-yielding seeds, wheat produced the best results. Production of coarse grains-the staple diet of the poor and pulses the main source of protein-lagged behind, resulting in reduced per capita availability.

Before the Green Revolution

The total area under the high-yielding-varieties program was a negligible 19,000 km² in FY 1960. Since then growth has been spectacular, increasing to nearly 154,000 km² by FY 1970, 431,000 km² by FY 1980, and 639,000 km² by FY 1990. The rate of growth decreased significantly in the late 1980s.

Nevertheless, when the British left India four years later in 1947, India continued to be haunted by memories of the Bengal Famine. It was therefore natural that food security was a paramount item on free India's agenda. This awareness led, on one hand, to the Green Revolution in India and, on the other, legislative measures to ensure that businessmen would never again be able to hoard food for reasons of profit.

Results

The major benefits of the Green Revolution were experienced mainly in northern and northwestern India between 1965 and the early 1980s; the program resulted in a substantial increase in the production of food grains, mainly wheat and rice. Food-grain yields continued to increase throughout the 1980s, but the dramatic changes in the years between 1965 and 1980 were not duplicated. By FY 1980, almost 75 percent of the total cropped area under wheat was sown with high-yielding varieties. For rice the comparable figure was 45 percent. In the 1980s, the area under high-yielding varieties continued to increase, but the rate of growth overall was slower. The eighth plan aimed at making high-yielding varieties available to the whole country and developing more productive strains of other crops.

The Green Revolution created wide regional and interstate disparities. The plan was implemented only in areas with assured supplies of water and the means to control it, large inputs of fertilizers, and adequate farm credit. These inputs were easily available in at least parts of the states of Punjab, Haryana, and western Uttar Pradesh; thus, yields increased most in these states. In other states, such as Andhra Pradesh and Tamil Nadu, in areas where these inputs were not assured, the results were limited or negligible, leading to considerable variation in crop yields within these states. The Green Revolution also increased income disparities: higher income growth and reduced incidence of poverty were found in the states where yields increased the most and lower income growth and little change in the incidence of poverty in other states.

AGRICULTURE (HIMACHAL)

Agriculture contributes nearly 45% to the net state domestic product. It is the main source of income as well as employment in Himachal. About 93% of the state population depend directly upon agriculture. However, agriculture in the state suffers from certain limitations, especially in the production of food grains. One of these reasons is that the area under cultivation can't be extended to an appreciable extent. Also, reclamation of land on slopes is not economical and increases environmental degradation. The state can profit more by cultivating cash crops as per the agro-climatic conditions.

The main cereals grown in the state are wheat, maize, rice and barley. Kangra, Mandi and the Paonta valley of Sirmaur district (to some extent) are the major producers of the first three cereals, while barley is mostly cultivated in Shimla.

Though the state is deficient in food grains, it has gained a lot in other spheres of agricultural production such as seed potato, ginger, vegetables, vegetable seeds, mushrooms, chicory seeds, hops, olives and fig. Seed potato is mostly grown in the Shimla, Kullu and Lahaul areas. Special efforts are being made to promote cultivation of crops like olives, figs, hops, mushrooms, flowers, pistachio nuts, sarda melon and saffron.

Fruit cultivation has also proved to be an economic boon. There are huge tracts of land suitable only for growing fruits. Fruit of all cultivation does not add to the problem of soil erosion and its employment potential is more than conventional farming. The yield per acre in terms of income is also much higher. Apple farming produces the maximum income. Fruit growing in the state is fetching over Rs.300 crore annually. The state has also earned the name of the *Apple State of India*.

2

Agriculture and Rural Developments

From a nation dependent on food imports to feed its population, India today is not only self—sufficient in grain production, but also has a substantial reserve. The progress made by agriculture in the last four decades has been one of the biggest success stories of free India. Agriculture and allied activities constitute the single largest contributor to the Gross Domestic Product, almost 33% of it. Agriculture is the means of livelihood of about two—thirds of the work force in the country.

This increase in agricultural production has been brought about by bringing additional area under cultivation, extension of irrigation facilities, the use of improved high yielding variety of seeds, better techniques evolved through agricultural research, water management, and plant protection through judicious use of fertilizers, pesticides and cropping practices.

CROPS

The 1970s saw a multi-fold increase in wheat production that heralded the Green Revolution. In the next decade rice production rose significantly; in 1995-96, rice production was 79.6 million tons. Total grain production crossed 191 million tons in 1994-95, a big leap from 51 million tons in 1950-5 1. During the Seventh Plan, the average grain production was 155 million tons, 17 million tons more than the Sixth Plan average.

To carry improved technologies to farmers, a National Pulse Development Program, covering 13 states, was launched in 1986.

The Special Food Production Program augmented efforts to boost pulse production further. In 1995-96, pulse production was 13.2 million tons. With some States offering more than the statutory minimum price, sugarcane production also received a boost, in 1995-96 a record 283.0 million tons was registered.

Irrigation

As efforts continued to increase the irrigation potential in the country, the last 40 years saw the gross irrigated area reach 8~ million hectares. Flood forecasting has become an important activity over the years. Over 500 hydrological stations collect and transmit data through 400 wireless stations for issuing forecasts for 157 sites. About 5000 forecasts are issued in a year with 94% accuracy. The country also receives international support, with the World Bank as a primary source, for developing the water resources. International cooperation is also envisaged in setting up a National Center for Information on Water and Power. As there is a broad seismic belt in the country, particularly along the Himalayan, the Kutch region and ports of Maharashtra, a scheme is being evolved to collect all data on seismic activity at various dam sites.

Fertilizers

The fertilizer industry in India has grown tremendously in the last 30 years. The Government is keen to see that fertilizer reaches the farmers in the remote and hilly areas. It has been decided to decontrol the prices, distribution and movement of phosphatic and potassic fertilizers. Steps have been taken to ensure an increase in the supply of non-chemical fertilizers at reasonable prices. There are 53 fertilizer quality control laboratories in the country. Since bio—fertilisers are regarded as an effective, cheap and renewable supplement to chemical fertilizers, the Government is implementing a National Project on Development and Use of Bio-fertilisers. Under this scheme, one national and six regional centers for organizing training, demonstrating programs and quality testing of bio-fertilisers has been taken up.

It was a challenging decision of the Government to take Bombay High gas through a 1,700-km pipeline to feed fertilizer plants located in the consumption centers of North India. However, the major policy which has ensured the growth of the fertilizer industry is the thrust on accelerating fertilizer consumption by fixing, on the one hand, low and uniform price for fertilizers, and on the other hand providing the manufacturers adequate compensation through the retention price and subsidy scheme. As expected, fertilizer nutrient demand has gone up from 0.29 million tons in 1960-61 to 13.9 million tons at the end of 1995-96, compared to 12.15MT during 1992-93.

Fisheries

Fish production achieved an all-time high of 4.9 million tons at the end of 1995-96. Programs that have helped boost production include the National Program of Developing Fish Seeds, Fish Farmers' Development Agencies and Brackish Water Fish Farmers' Development Agencies. The Central Institute of Fisheries Nautical and Engineering Training trains the necessary manpower. To diversify fishing methods and introduce processed fish products on a semi-commercial scale, an Integrated Fisheries Project has been launched. A National Fisheries Advisory Board has also been established.

Food Processing

A Ministry of Food Processing Industries was established in July 1988 to ensure better utilization of farmers' output by inducting modem technology into the processing of food products, thus augmenting the income of farmers and generating employment opportunities in rural areas. A new seeds policy has been adopted to provide access to high quality seeds and plant material for vegetables, fruit, flowers, oil-seeds and pulses, without in any way compromising quarantine conditions. Initiatives have been taken to encourage private sector investment in the food processing industry.

AGRICULTURAL RESEARCH

The apex body for education, research and extension education in the field of agriculture is the Indian Council of Agricultural Research (ICAR), established in 1929. India's transformation from a food deficit to a food surplus country is largely due to ICAR's smooth and rapid transfer of farm technology from the laboratory to the land.

ICAR discharges its responsibilities through 43 research institutes, four national research bureaus, 20 national research centers, nine project directorates, 70 all-India coordinated research projects, and 109 Krishi Vigyan Kendras (farm science centers). Besides, the program of Agricultural Education is coordinated by ICAR with the curricula and other normative guidance given to the 26 Agricultural Universities and four National Research Institutes.

Oilseeds Production

A Technology Mission on Oilseeds was launched in 1986 to increase production of oilseeds in the country and attain self-sufficiency. Pulses were brought under the Technology Mission in 1990. Before the Mission was launched in 1985-86, oilseed production was 10.83 million tons; during 1995-96, it was estimated at 22.42 million tons, which is a record. Soybean, rapeseed and mustard largely contributed the increase in production. Production of pulses has seen many ups and downs, which is expected to be checked under the Mission. The country grows mainly nine oilseeds, with groundnut, rapeseed and mustard accounting for 62% of total production. Lately, soybean and sunflower have shown major growth potential.

Drinking Water

A Technology Mission on Drinking Water and Related Water Management has been constituted to cover the residual problem villages and provide potable water at 40 liter per capita per day, and 70 liters per capita per day in desert areas inclusive of 30 liters for cattle. The Mission is tackling the problem through 55 mini-missions in project districts and countrywide problem oriented sub-missions. A Village Level Operation and Maintenance (VLOM) pump called India Mark-11 has been developed and is being exported to 40 countries. By March 31, 1993, over 79% of the rural and about 85% of the urban population was provided drinking water facilities.

AGRICULTURE, GROWTH AND POVERTY REDUCTION

History shows that different rates of poverty reduction over the past 40 years have been closely related to differences in agricultural performance – particularly the rate of growth of agricultural productivity.

In Asia, the rapid productivity gains of the Green Revolution increased producers' incomes, raised labourers' wages and lowered the price of food. In addition, new livelihood opportunities were generated when success in agriculture provided the basis for economic diversification. However, despite decades of investment in new agricultural technology and rural development, hunger and poverty continue to plague large areas of the developing world. The problem is particularly acute in sub-Saharan Africa, where progress towards the Millennium Development Goals is slowest.

While increasing agricultural productivity perhaps remains the single most important determinant of economic growth and poverty reduction, serious doubts are emerging as to whether agricultural productivity can be further increased, where it is most needed, and what part, if any, small-scale farming will play in the future. Development experts need greater understanding of the links between agricultural productivity and poverty. They also need to assess just how far they have changed and the extent to which small-scale agriculture can remain a ladder out of poverty for millions of poor people living in rural areas.

Links between agriculture and poverty reduction are forged through four 'transmission mechanisms':

- direct and relatively immediate impact of improved agricultural performance on rural incomes;
- impact of cheaper food for both urban and rural poor;
- agriculture's contribution to growth and the generation of economic opportunity in the non-farm sector; and
- agriculture's fundamental role in stimulating and sustaining economic transition, as countries (and poor people's livelihoods) shift away from being primarily agricultural towards a broader base of manufacturing and services.

However, the potential for future poverty reduction through these transmission mechanisms depends on the extent to which agricultural productivity can be increased where it is most needed or, more simply and directly, the extent to which the Green

Revolution can be replicated in Africa. Whether and how this might be achieved has been the subject of protracted debate amongst researchers and policy makers. The main questions in this debate are:

- Where to focus development efforts – high potential areas where development options are greater or poorer areas where the potential and options are less but the incidence of poverty is greatest?
- Who to focus on – the poorest smallholders who produce for subsistence and have limited engagement with markets, or larger-scale farmers whose success can be an engine of growth and generate wealth and jobs for rural households?
- What to do in less favoured areas – should agriculture be promoted amongst poor households in areas of low agricultural potential that are remote from markets and inputs supplies? Or should the focus be on employment generation to enable poor households to buy in food?
- What role for technology – should primacy in development efforts be given to yield-raising technology or to less intensive approaches that minimise variation?
- Which crops – should the emphasis be on crops that will be largely consumed within the households, or on income-generating cash crops?

Contrasting answers to these questions produce different policy positions. Those who are optimistic about the role of smallholders emphasise the need for increased direct investment in agriculture and rural development, with support focusing on creating institutions that will encourage and support smallholder-led agricultural development. They believe it is important to improve the productivity of staple food crops that are not traded internationally, but consumed by the poor and traded locally.

In contrast, smallholder pessimists emphasise the need to achieve the best outcome possible from rapidly changing global markets for agricultural produce. They believe that commercial production (probably on larger farms) of non-staple cash crops should be encouraged, particularly those that result in robust links to the non-farm sector where the main source of employment for the rural poor would be found. Influencing international policy processes to ensure access to developed country markets is a priority for this position, as is improving the human capital assets of the rural poor so they can take up opportunities in growth areas as they occur.

What is the Issue?

Poverty has fallen rapidly over the past 40 years, but at different rates around the world. Asia has achieved the most rapid poverty reduction, particularly China, but also India and South East Asia. In contrast, little if any progress has been made in sub-Saharan Africa, where the number of people living on less than one dollar a day – the internationally agreed definition of absolute poverty – has doubled over the past 20 years (World Bank, 2004a).

Historically, rates of poverty reduction have been very closely related to agricultural performance – particularly to the rate of growth of agricultural productivity. In simple terms, this indicates that the countries that have increased their agricultural productivity the most have also achieved the greatest reductions in poverty.

In Asia rapid productivity gains, achieved largely through the technological advances of the Green Revolution, provided a 'fast-track' route out of poverty by directly increasing producers' incomes and labourers' wages, by lowering the price of food and by generating new livelihood opportunities as success in agriculture provided the basis for economic diversification.

However, despite decades of investment in new agricultural technology and rural development, hunger and poverty continue to plague large areas of the developing world.

The problem is particularly acute where people depend on rain-fed agriculture, in particular sub-Saharan Africa, where the impact of new technologies has been less apparent and agricultural productivity has generally stagnated and even fallen in some areas.

Achieving the Millennium Development Goal (MDG) of halving the proportion of people living in absolute poverty by 2015 will depend largely on increasing agricultural productivity, which remains perhaps the single most important determinant of economic growth and poverty reduction. This fact is not lost on developing countries or their development agency partners, who are seeking ways to stimulate agricultural development. But serious doubts are emerging as to whether agricultural productivity can be increased where it is needed most, and what part, if any, small-scale farming will play in the future. Underlying such doubts is a concern that the context in which small-scale agriculture could achieve productivity gains today is very different to that which prevailed in Asia during the halcyon days of the Green Revolution. Inherent differences in production capabilities and quite fundamental changes in the international agricultural context are combining to create a set of circumstances that are probably far less conducive to achieving the type of productivity transformation witnessed in Asia.

Development experts need greater understanding of the links between agricultural productivity and poverty. They also need to assess just how far they have changed and the extent to which small-scale agriculture can remain a ladder out of poverty for millions of poor people living in rural areas.

AGRICULTURE, GROWTH AND POVERTY – WHAT WE KNOW OF THE RELATIONSHIP

The Context – the State of World Poverty

Between 1981 and 2001, the percentage of the world's population living on less than a dollar a day fell from 40.4% to 21.1%. Even though world population grew by an estimated 1.5 billion over the same period, the numbers of people living in absolute poverty fell by almost 400 million to around 1.1 billion. These figures, while still presenting a picture of massive global poverty, represent probably the fastest rate of poverty reduction ever witnessed. However, the aggregate picture masks a story of variable progress (Table 1).

In terms of the proportion of people in absolute poverty, significant reductions occurred in most regions between 1981 and 2001. Progress was most rapid in East Asia, primarily China, where the proportion of the population living in absolute poverty decreased from almost 64% to just 16.6%. Elsewhere progress was steady, with the stark exception of sub-Saharan Africa where the percentage of the population living on less than US$1 a day has increased from 41% to almost 47%.

Table 1. Percentage and numbers of population living below the US$1 per day poverty line, 1981–2001

Region	% of population living below US$1			Number of people living on less than US$1		
	1981	1990	2001	1981	1990	2001
East Asia	57.7	29.6	14.9	795.6	472.2	271.3
China	63.8	33.0	16.6	633.7	374.8	211.6
Europe and	0.7	0.5	3.7	3.1	2.3	17.6
Latin Americ a and	9.7	11.3	9.5	35.6	49.3	49.8
Middl e East and	5.1	2.3	2.4	9.1	5.5	7.1
South	51.5	41.3	31.3	474.8	462.3	428.4
India	54.4	42.1	34.7	382.4	357.4	358.6
Sub-Saharan	41.6	44.6	46.9	163.6	226.8	315.8
Global						
	40.4	27.9	21.1	1481.8	1218.5	1092.7

PPP = ??? Source: World Bank (2004a)

In terms of the numbers of people living in absolute poverty, the picture is very different. Almost the entire reduction in numbers between 1981 and 2001 can be attributed to progress in China, along with generally small and disappointing reductions in other parts of East Asia, India (where over 350 million people still live in absolute poverty) and the Middle East. But in sub-Saharan Africa, the number of people in absolute poverty has effectively doubled.

Almost all of these changes (both positive and negative) can be explained by a variable performance in tackling persistent rural poverty. Changes in rural poverty figures tend to closely track changes in national poverty figures, primarily because poverty in the developing world is a largely rural problem. As an illustration, the national poverty headcount in India decreased from 36% to 29% between 1994 and 2000, while rural poverty decreased from 37% to 30%.

Moving away from the broad geographical distribution of poverty described above, the characteristics of poverty remain remarkably consistent:

- poverty remains a predominantly rural problem: 75% of the world's poor live in rural areas (IFAD, 2001);
- poverty is concentrated amongst women, children and the elderly;
- poverty is most concentrated amongst vulnerable groups and castes, landless people and ethnic minorities), who often live in remote rural areas. In Vietnam, for instance, the national poverty headcount fell from 58% to 29% during 1993–2002 for the whole country but remained at 68% in the northwestern hill areas, which are predominantly inhabited by ethnic minorities (World Bank, 2004).

As the next two sections of this paper will explore, differences in agricultural performance – particularly different experiences in increasing agricultural productivity – go far in explaining these major differences in reducing poverty.

Agriculture's Recent Performance – a Picture of Mixed Progress

Total agricultural production has increased rapidly during the past 40 years. Statistics from the Food and Agriculture Organisation of the United Nations (FAO, 2004) indicate that between 1961 and 2001, global cereal production more than doubled (from 900 to 2,100 million tonnes), far outstripping the rate of growth of population. In the developing world as a whole, per capita food production rose by over 50% between 1961 and 2001 (IFPRI, 2002a). This has ensured that world food prices have declined progressively in real terms; Mitchell and Ingco (1993) report a 78% fall in food prices overall from 1950 to 1992. Some of this increase reflects an expansion of the area under agriculture – particularly in Africa – but global yields have also increased over the same period, reflecting an increase in productivity through the introduction of new technologies and a major expansion of irrigation. (It is important to note the fundamental difference between increasing production and increasing productivity). The generally positive global picture masks large and quite fundamental regional variations. These are mapped out in Figures 1 and 2 and described below.

Production and Productivity

To understand the sources of growth, it is important to distinguish between production and productivity.

Production is the same as output. It is physical produce and can be reported in units of volume or weight. For instance, cereal production would be reported in metric tonnes.

Productivity is not physical produce; it is a number. Productivity is defined as output per unit of input, where 'input' can be land, labour and/or capital, and 'output' is agricultural produce. The importance of productivity, however precisely defined, is that it gives a measure for efficiency. It tells us in one figure how much input was used to produce a unit of output. For instance, the labour productivity of paddy rice in 1998 in India is a number that tells us the amount of paddy rice that one Indian agricultural worker produced in 1998, on average. The unit of labour productivity is kilograms per worker.

It is important to specify what output (or outputs) and what input (or inputs) are actually used to calculate productivity. Labour productivity is the volume of output per unit of labour per worker (e.g. kg/day/person). Land productivity is the volume of output per land unit (e.g. tonnes per hectare). Both labour productivity and land productivity can be calculated for a single product or for a set of products. Finally, total factor productivity is the volume of all outputs divided by the volume of all inputs. Here the various inputs and outputs are added up to obtain one quantity of input and one quantity of output. These are divided (output/input) to obtain a figure for total factor productivity.

Agricultural performance in Asia over the past 40 years has been remarkable. Between 1961 and 2001, cereal production in the region's developing countries increased from 309 to 962 million tonnes, a rate of increase far in excess of population growth. In the same period, the expansion of agricultural land was comparatively modest (from 1.0 to 1.4 billion hectares). The growth in output came mostly from increasing agricultural productivity. For instance, cereals productivity nearly tripled from 1.2 to 3.3 tonnes/ha between 1961 and 2001 in developing Asia (all figures from FAO, 2004). This rate of productivity growth has had an enormous impact on economic development and poverty reduction.

In contrast, over the same period, the production of basic cereals in sub-Saharan Africa grew only fourfold, from 40 to 116 million tonnes, but the increase in production barely kept pace with population growth. Furthermore, productivity increases were very small. Thus, while cereals productivity in Africa rose by a half (from 0.8 to 1.2 tonnes/ha), most of the output growth came from using more land and labour, not more productive use of those resources. The area of agricultural land in Africa expanded from one million to one billion hectares during 1961–2001 (all figures from FAO, 2004). Weak productivity growth in agriculture had serious consequences for economic development and poverty reduction. The next section of this paper considers why agricultural productivity is important to poverty reduction.

Agricultural Growth and Poverty Reduction – the Evidence

At the macro-economic level, growth in agriculture has been consistently shown to be more beneficial to the poor than growth in other sectors. Furthermore, analysis reveals that increasing agricultural productivity has probably been the single most important factor in determining the speed and extent of poverty reduction during the past 40 years. Much of this evidence is derived from the Green Revolution in Asia – examples from Africa are noticeably fewer.

With respect to the pro-poor benefits of growth in agriculture, Datt and Ravallion (1996) showed that rural sector growth in India reduced poverty in both rural and urban areas, while economic growth in urban areas did little to reduce rural poverty. Warr

(2001) provided evidence that growth in agriculture in a number of South East Asian countries significantly reduced poverty, but this was not matched by growth in manufacturing.

Gallup et al. (1997) showed that every 1% growth in per capita agricultural Gross Domestic Product (GDP) led to 1.61% growth in the incomes of the poorest 20% of the population – much greater than the impact of similar increases in the manufacturing or service sectors. Numerous other studies reveal similar results, but emphasise the important qualification that the degree to which agricultural growth reduces poverty is usually conditional upon the initial distribution of assets (in particular land) and the initial level of inequality (Bourgignon and Morrison, 1998; Timmer, 1997; de Janvry and Saddoulet, 1996).

In terms of the role of agricultural productivity in reducing poverty, Thirtle et al. (2001) concluded from cross-country regression analysis that, on average, every 1% increase in labour productivity in agriculture reduced the number of people living on less than a dollar a day by between 0.6 and 1.2%. No other sector of the economy shows such a strong correlation between productivity gains and poverty reduction. The routes through which growth in agriculture achieve such a potent impact on poverty are considered below.

Understanding how increased agricultural productivity reduces poverty

Four 'transmission mechanisms' critically link changes in agricultural performance, more especially productivity increases, to progress in reducing poverty:

- direct and relatively immediate impact of improved agricultural performance on rural incomes;
- impact of cheaper food for both urban and rural poor;
- agriculture's contribution to growth and the generation of economic opportunity in the non-farm sector; and
- agriculture's fundamental role in stimulating and sustaining economic transition, as countries (and poor people's livelihoods) shift away from being primarily agricultural towards a broader base of manufacturing and services.

The direct impact on rural poverty

Poverty remains a predominantly rural problem and agriculture is generally central to rural livelihoods. Some 70% of the workforce in sub-Saharan Africa and 67% in South Asia are at least partly engaged in agriculture (Maxwell, 2001). Therefore, any improvement in rural incomes should – if only by sheer weight of numbers – have a major impact on poverty.

The most useful assessments of the impact on poverty of changes in agriculture are those that follow farming communities' experiences over a long-term period (Lanjouw and Stern, 1998; Hazell and Ramasamy, 1991). These studies showed that agricultural productivity gains have raised rural incomes in two ways: by directly increasing farmers' incomes and, of particular importance to the poorest, by increasing employment opportunities and wages.

Farmers' incomes

De Janvry and Sadoulet (1996) estimate that in Asia, a 10% increase in total factor productivity in agriculture would raise the incomes of small-scale farmers by 5%. Acharya and Sophal (2002) report that in a 2001 sample of smallholder rice-producing farms in Cambodia, a 10% increase in yields resulted in an 8.8% increase in household incomes in dry season cultivation and a 4.4 % increase in wet season cultivation. Lipton and Longhurst (1989) and Hazell and Ramasamy (1991) provide similar evidence.

Two further observations on this issue deserve a mention. First, farm incomes have generally continued to rise despite declining market prices resulting from major output expansion. By adopting new technologies and expanding irrigation, farmers have been able to progressively reduce their unit costs of production and so remain profitable. Bangladesh provides an excellent example. Between 1980 and 2000, the real wholesale price of rice in Dhaka fell from 20 Taka to 11 Taka per kg, but over the same period, farmers increased yields from around 2 to 3.4 tonnes per hectare, effectively offsetting the impact of falling

prices on their incomes. Second, there is no systematic evidence of smallholders being excluded from technology-led productivity gains. Lele and Agarwal (1989) cite evidence from

Kenya, where small-and large-scale farmers exist alongside one another, grow the same crops and sell them in the same markets at similar prices. Rohrbach and Makhwaye (1999) report that in Botswana a high-yielding sorghum variety released in 1994 had been adopted by almost 50% of the nation's small-scale farmers, who had planted it within two years of its release. In some exemplary Green Revolution countries, the numerical importance of small farms in agriculture actually increased during the technological transformation of agriculture. The International Rice Research Institute (IRRI) reports that in India the proportion of land holdings of less than one hectare increased between 1960 and 1990 from 41% to 59% of the total number of holdings (IRRI, 2004).

Employment

On-farm employment is critically important to poor people's livelihoods, and not just for the landless – agricultural labouring is a key means for many farmers to supplement their incomes. Evidence on this subject is primarily drawn from the Green Revolution experience in Asia. In India increasing agricultural productivity associated with the adoption of new technologies clearly increased demand for labour. Furthermore, and of particular benefit to the poor, the majority of the additional labour used was hired rather than family labour (Lipton and Longhurst, 1989; Hazell and Ramasamy, 1991). While intensification may involve some labour-economising innovations – particularly mechanical threshing – new varieties and irrigation have allowed farmers to double-and even triple-crop the land, which has consistently increased labour demand (Binswanger, 1986).

Significant increases in agricultural wage rates have been recorded in many countries. Saxena and Farrington (2003) showed that agricultural labour wages in India rose at a rate of about 3% per annum during the 1970s and 1980s.

The impact of reduced food prices

From the mid-1960s, when Green Revolution technologies began to be adopted widely, increases in the production of staple foods in most developing countries have comfortably outstripped population growth. Only in sub-Saharan Africa have food supplies grown slower than the population during the past 40 years.

Given this significant increase in per capita supply, and the relatively low elasticity of demand for basic foods, the real world market prices of the major traded grains have been in near continuous decline since the early 1950s. At the individual country level, increased production of food grains has often had a dramatic effect on reducing prices. This is of great benefit to the poor, both in urban and rural areas where many people buy and grow their own food.

Once again, Bangladesh is a good example. Between 1980 and 2000, production of rice and wheat increased from less than 15 to over 25 million tones, increasing per capita availability from 425 to 510 grams per day, despite population increasing over the same

period from 90 to 191 million. Real wholesale prices of rice and wheat in Dhaka have consequently fallen dramatically. Thus, poor people in urban areas (and net consumers of wheat and rice in rural areas) need a smaller proportion of their incomes to meet their basic food requirements.

Agriculture's contribution to growth and the creation of economic opportunity in the non-farm sector

Agriculture remains the economic heart of most developing countries. In Africa it provides two thirds of employment, generates over one third of Gross National Income (GNI) and over half of export earnings. In Asia, where economic growth and diversification have been most rapid, agriculture still provides jobs for 60% of the working population and 27% of GNI. Given agriculture's relative dominance in the economy, it remains the most likely source of significant growth in most developing countries. More immediately, fluctuations in agricultural performance are felt quickly, not only in the agricultural sector, but also in the wider non-farm economy.

Empirical work clearly illustrates the importance of these links. Timmer's (2003) analysis of the Kenyan economy showed that between 1987 and 2001, the rate of growth of the non-agricultural sector depended strongly on growth in agriculture. Non-agricultural growth increased by 30% of the agricultural growth rate in the same year, and by 10% of the agricultural growth rate in the previous year. Stern (1996) found a similar and significant relationship between growth in the agricultural and non-agricultural sectors during 1965– 1980 for a large number of developing countries.

The extent to which changes in agricultural performance influence the wider economy is determined by the size of the 'multiplier'. This is a measure of the extent to which a unit change in income earned in agriculture causes a change in income in the non-farm sector. The size of the multiplier will vary between places and over time, reflecting differences and changes in factors such as the amount of farm income spent on imported goods or saved. Thirtle et al. (2001) presents evidence from a number of studies and found multipliers ranging from 1.35 to 4.62.

These studies illustrate that agriculture – given its size and the impact of multipliers – is the most crucial sector for promoting broader economic growth and has greater 'knock-on' effects than any other sector. The case of Tamil Nadu (Box 2) is particularly interesting in this regard.

It has been argued that multiplier links in Africa are lower than in other parts of the world, and this has important consequences for wider economic development. However, Delgado et al. (1994) suggest a more optimistic estimate, with multipliers in excess of 1, indicating that the sector has considerable potential for stimulating broader economic growth.

AGRICULTURAL DEVELOPMENT AND ITS LINKS IN TAMIL NADU, INDIA

In North Arcot District, Tamil Nadu, 11 villages were surveyed in the early 1970s at the start of the Green Revolution, and again in the early 1980s. During this time, there had been almost complete adoption of high-yielding rice varieties, with much greater use of fertiliser and irrigation resulting in a modest, but sustained 60% increase in rice output between 1963 and 1980.

However, even more remarkable was that the poor were about twice as well off in the early 1980s as in the early 1970s. Real wage rates rose by 20% for men and 10% for women. This was not due to increased labour demand in farming (labour use fell as mechanisation cut jobs). In fact, members of farming households with increased incomes no longer offered themselves as casual labour, so the labour supply reduced at the same time. The main point, however, was the strength of growth links. For every rupee generated in increased farm output, R 1.87 was created in the off-farm economy, with about half in the demand for inputs, marketing and processing of crops, and half in meeting consumer demand. Source: Hazell and Ramasamy (1991).

Stimulating and Sustaining Economic Transition

The relative decline of agriculture's importance to the economy and people's livelihoods is an inevitable and desirable consequence of successful economic development. Rapid economic growth, sustained poverty reduction and substantial wealth creation have historically resulted from economic diversification, with less reliance on agriculture and more on manufacturing and services. However, history shows that most countries cannot successfully industrialise without with first achieving significant improvements in agricultural performance – particularly increased agricultural productivity. This holds true for Europe, North America, Japan, and the emerging countries of Asia where industrialisation has been very clearly agriculturally led (Timmer, 1988).

Understanding how growth in agriculture first stimulates and then sustains industrialisation has been the focus of much development thinking during the past 50 years. There is now a good theoretical understanding of how 'agricultural transformation' occurs, and a body of empirical evidence showing its effects. These are described in Lewis (1954) and Johnston and Mellor (1961) and summarised in Timmer (1988). The studies emphasise a number of key functions provided only by growth in agriculture and that enable broader economic growth and development to occur:

- generation of additional demand for goods and services produced outside of the agricultural sector as agricultural-based incomes rise – the size of agriculture and its multiplier effects is critically important here;
- generation of savings through increased farm incomes which can then be invested;
- provision of an available labour force;

- provision of affordable food which allows urban areas to develop and maintain wages rates at competitive levels; and
- provision of a raw material base to support manufacturing – processing of agricultural commodities has often been the first activity to be industrialised in many countries.

Two factors are of central importance to understanding agricultural transformation. Increased productivity (as opposed to increased production) is one key. Broad-based growth and diversification do not happen when agricultural output increases simply by using additional land or labour. Instead, greater value must be added to the land and labour used, i.e. agricultural productivity increases. This is where the historical contrast between Asia and Africa is relevant.

The second factor is that there is a paradox at the heart of agricultural transformation. Investment and growth must begin in agriculture so that the wider economy can go on to outgrow it. There is nothing inherently wrong or undesirable in this transition – quite the contrary – but agriculture must grow rapidly before the transition can occur. Growth and poverty reduction strategies that aim to bypass agriculture will almost certainly fail and will probably leave the population and economy locked in low productivity agriculture and poverty.

Emerging Issues and Questions

Can agriculture still provide the key to poverty reduction?

Asia's progress in freeing millions from poverty over the past 40 years can be largely attributed to the region's success in increasing agricultural productivity. It was also important that wider circumstances, policies and measures allowed agricultural productivity to occur, and ensured its impact on the wider economy. However, the Asian experience

The term 'agricultural productivity' is here used without specifying whether this is labour productivity or land productivity. During successful agricultural transformation, both typically increase, and their relative increases determine how much incomes and employment will rise' stands in sharp contrast to Africa, where agricultural productivity has been stagnant, rates of economic growth disappointing and poverty is on the rise.

Given the above, it is easy to reach the conclusion that without a significant improvement in agricultural performance – specifically improving agricultural productivity – the outlook for growth and poverty reduction in Africa remains poor. But Africa does not have a monopoly on poverty – the largest numbers of poor people live in rural Asia. While many live in remote areas or those that are weakly integrated into the wider economy, many also live in places that have witnessed the full effect of the Green Revolution, but its impact on their poverty has been limited.

Few doubt that achieving the MDG of halving the number of people living in absolute poverty by 2015 will require a significant improvement in agricultural performance, particularly in Africa. But in looking at the future and the likelihood of this being achieved, differences of opinion emerge around two key questions:

1. Do the conditions exist for agricultural productivity to be increased where it is most needed and what part, if any, can small-scale agriculture play in achieving this?
2. Given quite fundamental differences in context between Asia in the Green Revolution and today's poorest countries, will the historical relationship between agricultural growth and poverty reduction continue to hold true?

THE VIEW TAKEN ON BOTH THESE QUESTIONS WILL HAVE MAJOR POLICY IMPLICATIONS.

Are significant improvements in agricultural productivity possible?

Increasing the pace of poverty reduction – most critically in Africa – will depend upon the extent to which agricultural productivity can be increased through a step change in agricultural performance. Simply increasing output at current productivity levels – and even this appears difficult in some places – will have little long-term impact on growth or poverty. While agreement exists on the need to increase productivity, opinion is divided on whether rapid productivity gains, similar to those seen in Asia during the Green Revolution, can be achieved elsewhere.

Much of this debate focuses on whether the economic and physical preconditions that enabled the Green Revolution to happen exist in those regions as yet untouched by this kind of transformation. Inevitably this debate seeks to contrast Africa's failure with Asia's success, but such a stark characterisation of debate should be approached cautiously. There have been incidences of rapid productivity improvement in Africa (Wiggins, 2000) and the lives of millions of poor people in Asia remain largely unchanged despite the Green Revolution (Rosset et al., 2000). However, a qualified comparison of Asia in the 1960s–1970s and sub-Saharan Africa today provides a useful basis to explore these differences.

Differences in the Physical Resource Base—Weather, Water and Soil

Africa's physical conditions are generally less conducive to improving agricultural productivity than those in Asia. Weather patterns are more unpredictable and soils generally less productive, making agriculture a risky venture. Farmers are often reluctant to adopt new and untried technologies because of the high cost of failure.

Availability of water resources is probably a major factor. Over the past 40 years, productivity gains achieved in irrigated areas have consistently outpaced those of rain-fed areas. For instance, maximum attainable land productivity in Indian rain-fed paddy rice production is 2,516 kg/ha compared to 8,161 kg/ha in irrigated paddy rice production.

For wheat production, the respective numbers are 1,786 kg/ha and 4,352 kg/ha (FAO, 2003). Asia's success in realising the benefits of high-yielding varieties depended critically upon a massive expansion in irrigated area, from 40 million hectares in 1961–1963 to 71 million hectares in 1997–1999. In contrast, only 5 million hectares are currently irrigated in sub-Saharan Africa (most of it in South Africa), a small change from the figure of 3 million hectares irrigated in 1961–1963 (FAO, 2003). While some further increase in irrigation in Africa could be achieved, it is questionable whether Africa's water resources could support expansion on the scale seen in Asia.

Differences in Population and Infrastructure Density

Here the argument is that Africa's population density has not yet reached the level that characterised Asia 40 years ago, and that allowed the efficient provision of rural infrastructure, such as roads and marketing facilities, in Green Revolution adopters. In the absence of such infrastructural support, the probability of small-scale farmers being able to move into semi-commercial production, which would give them the finance and incentive to purchase inputs, is much lower. Provision of rural infrastructure varies greatly in Africa, but even today it remains far below the level seen in Asia 30–40 years ago. Road density is crucial for agricultural intensificationt. Yet road density in Africa is only 34 miles per square kilometre, compared to 500 in India (Riverson et al., 1991).

Declining Real Prices and Adverse Terms of Trade

The greatest advance of the Green Revolution took place during the 1960s, within a particular international price environment. Since then agricultural commodity prices have more than halved in real terms as global output has expanded. Although the rate of decline appears to have slowed down since the 1980s, the question remains – are prices sufficiently remunerative to encourage farmers to innovate and take risks? Evidence has been found to support both sides.

Pessimists are particularly concerned over adverse movements in the maize fertiliser price ratio, and its implications for increasing fertiliser use, which is the key to raising yields and productivity. They note that while the real price of fertiliser has fallen, the decline in real food prices has been even greater. Optimists note that the link between international and local prices is often quite weak. Domestic markets, particularly in rural Africa, are poorly integrated into regional and international markets and high transport and distribution costs have, and may well continue to, offset at least some of the downward pressure on local prices caused by falling international prices. This is clearly illustrated in countries where the main staple is imported from neighbouring producers. Here domestic food prices are often more influenced by movements in fuel prices than changes in international commodity prices. The impact of liberalisation on agricultural trade will have an uncertain bearing on the prices producers will realise. Accession to the World Trade Organisation (WTO) has required several countries to open up their domestic markets to competition. The impact has varied from country to country, from crop to crop and even within countries. Estimates by FAO and the United Nations Conference on Trade and Development (UNCTAD) on the impact of liberalisation on global prices have, in general, predicted an upward impact on world commodity prices, primarily as developed country production declines with the removal of farm subsidies.

Analyses of specific country experiences have shown more mixed outcomes. For example, a recent Oxfam study (2002) suggests that a 10% reduction in import tariffs on rice in Senegal resulted in a doubling of rice imports from Thailand. But even within developing countries, liberalisation will have mixed benefits tending to favour more outward-looking regions with better infrastructure and more competitive production systems. A study by Chen and Ravallion (2003) of the impact of China's accession to WTO shows that the richest rural regions benefited most, while less well developed regions suffered from falling prices due to increased openness. Predicting the producer price impact of liberalisation will be difficult, but it is unlikely to benefit the poorest agricultural producers and it seems likely that small-scale farmers will face more adverse terms of trade than their predecessors.

The Impact of HIV/AIDS

A serious issue of growing importance is the impact of HIV/AIDS on agriculture. The epidemic is already having far-reaching effects on rural households and communities in sub-Saharan Africa. In Asia, although prevalence rates generally remain low, they are increasing rapidly. HIV/AIDS affects agricultural production through both labour and capital effects. Mortality and morbidity effects reduce the availability of labour, whilst the cost of treating and eventually burying those who develop AIDS diverts investments away from agriculture. Since labour is one of the most important assets that poor people have, significant improvements in agricultural productivity will be compromised by HIV/AIDS. It will become more difficult to develop labour-intensive approaches as a strategy for reducing poverty. In certain regions and countries within Africa and Asia, it is possible that a combination of urban migration and HIV/AIDS may further exacerbate labour shortages, although this could drive up agricultural wage rates.

New Barriers to Trade—the Impact of Emerging New Standards

Globalisation, as characterised by the continued expansion in world trade, increasing interdependence of world markets and the internationalisation of world production, presents additional challenges to developing countries. One new aspect is the emergence of standards imposed by national and regional bodies (such as the European Union) for reasons such as food safety (Wilson and Abiola, 2003). Nationally and internationally imposed food safety and phytosanitary standards represent a new block to export markets for large numbers of developing country producers.

THE IMPACT OF CHANGES IN DOMESTIC SUPPLY CHAINS

The impact of globalisation also extends to domestic markets. Changes in supply chains, in particular the increasing presence of large supermarket chains in many developing countries, are creating new demands on quality, quantity and timeliness of products. To meet these new demands, agriculture is becoming increasingly industrialised and capital intensive, making it harder for individual family farmers to access markets (Barrientos and Kitzinger, 2003). The most marked change has occurred in Latin America, but it is also increasingly evident in Africa. IFPRI (2004) estimates that supermarkets now account for up to 30% of food retailing in Kenya. The concern for many is that globalisation is offering opportunities to the rural poor, but as low-paid employees, rather than as farmers.

A Changing Policy Context

A common characteristic of successful Green Revolution adopters was the primacy awarded to agriculture in national development efforts, and the role played by the state in supporting agriculture. From the 1950s onwards, Asian governments provided (and often subsidised) key inputs, constructed infrastructure (notably irrigation) and engaged in product markets to ensure stable, predictable and remunerative prices. Through such instruments, governments created a lower risk environment for agricultural innovation and increased its affordability for small-scale farmers with considerable success. In Africa, as Wiggins (2000) notes, occasional episodes of rapid agricultural development over the past 30 years have occurred when the state undertook a similar progressive role. In both instances, the approach was supported by the international development community.

Whether the provision of such public services is affordable or can be managed effectively by the state (and the evidence points to this being problematic), are matters for debate. First, it is clear that shifts in international views, particularly those of the international financial institutions, on which most of the poorest countries depend, now rule out many of the policy interventions that assisted innovation and rapid productivity growth in Asia (Dorward et al., 2004). Just how far this has negatively influenced the diffusion of new technological innovation and slowed progress in increasing productivity is a moot point. Second, whilst evidence about the precise impacts on agricultural production and agriculture-based livelihoods is very patchy, there may be policy implications associated with state involvement in driving production (in the face of reduced labour capacity) and people's access to food. Governments in Africa may be facing long-term welfare bills to support both food availability and access.

What are the Best Agricultural Development Strategies?

There is probably less of a consensus now – particularly amongst development agencies – on the best (in terms of impact on poverty and hunger) agricultural development strategy than at any time over the last half-century or longer (Ashley and Maxwell, 2001). This is particularly true of Africa, where an unsuccessful model based on improving performance through technology supported by publicly owned development agencies has been replaced by the equally disappointing response of farmers to the liberalisation of markets. Key points in the debate are considered below.

Where should development efforts be focused to achieve the greatest return in terms of reducing poverty and hunger? Should they be focused on high-potential areas where development options are greatest, or in the poorest areas where there are fewer potentials and options, but where poverty is greatest?

Who should they be focused on? Here the debate sees at least three positions:

1. Accept the demise of the peasantry and work with large-scale farmers whose success will act as a catalyst to generate wealth and jobs for those whose farms are not viable. Proponents of this view (see Maxwell, 2004) identify changes in global supply chains as being major new obstacles to smallholders that will prove insuperable for many or most.
2. Work with smallholders, but accept that most innovation, investment and commercialisation will come from only that (possibly very small) portion with more land and capital than their neighbours. Some claim that these farmers will then create enough jobs locally, through hiring labour and spending on local goods and services, to boost the welfare of other farm households (David et al., 2000).
3. Focus on the poorest and most disadvantaged smallholders to beat poverty and hunger and reduce vulnerability directly (IFPRI, 2002a and 2002b).

Should they focus on less favoured areas? These include poor households in areas of low agricultural potential that are remote from markets and supplies of inputs. There are two positions on this question:

1. In remote areas, employment opportunities in the rural non-farm economy are often limited. Thus, in spite of poor prospects in farming, people are heavily dependent on crops and livestock for their livelihoods. The promotion of and investment in agriculture should therefore be viewed as a safety net provision in itself, irrespective of whether such agriculture is contributing to growth.
2. In many of these cases, food security will be assured more by the ability to buy in food, rather than by trying to

produce more. The questions posed for such areas are those of jobs and incomes. The difficulty lies in trying to create jobs where resources and infrastructure are scarce and markets remote. The answer probably lies in a combination of marginal agriculture, forestry, fishing, tourism, public employment in provision of services and physical infrastructure (and its maintenance), public transfers for social protection and regional equity, and in migration to alternative opportunities (Hite, 1997). Agricultural development may not, in these areas, be a prime mover in reducing poverty and improving food security.

What is the role of technology? Should development efforts focus mainly on yield-raising technology or on less intensive approaches that minimise variation? Malawi is a good case in point – which is the better option: high-yielding hybrid maize with fertiliser applications or lower-yielding open-pollinated varieties requiring less fertiliser?

Which crops? Should the accent be on crops that will be largely consumed within the household or on income-generating cash crops?

Two Contrasting Views on Agricultural Strategies

Most analysts agree that the context for agricultural development has changed significantly over the last 40 or so years, and is probably now less conducive to the type of smallholder-led agricultural development seen in Asia in the 1960s and 1970s. However, opinion is sharply divided on what these changes really mean for agriculture's role in poverty reduction and, in particular, what they imply for small-scale farming as a route out of poverty for millions of poor people. Two broad views exist: the smallholder optimists and the smallholder pessimists. Each offers a different interpretation of the impact of the changes outlined above on the role of agriculture in poverty reduction and suggests a very different development strategy for agriculture.

The Smallholder Pessimists

Pessimists (e.g. Maxwell, 2004) argue that with the changes described above, smallholder agriculture is becoming progressively less viable in many parts of the world. If these trends continue, the point will soon be reached where small-scale agriculture is so uncompetitive that it ceases to offer a direct route out of poverty for the rural poor.

This vision is consistent in many ways with the way agriculture has progressed in the developed world. It sees agriculture based around farmers who can operate within the framework of new conditions, with consolidation of small farms into larger commercial units being an inevitable consequence. While this process took generations to occur in the developed world, the pace of globalisation is such that it is occurring far more rapidly in the developing world today. Policy makers need to be aware of the possible outcomes.

As smallholder agriculture declines and rural areas become better connected to urban centres, offering new opportunities, the future for rural people may lie less in agriculture and more in other economic sectors. Pessimists note that this is already starting to happen in some places. Rural people are already diversifying their livelihoods away from agriculture and, for many, income from the non-farm rural economy, remittances, seasonal migration and even commuting are more important than that from agriculture (Lanjouw and Lanjouw, 2001; Start, 2001). Pessimists argue that while agriculture was important to poverty reduction in the past and remains an activity conducted in rural areas, this does not mean that it will automatically remain the primary means of earning a livelihood for millions of rural poor, nor does it represent an effective route out of poverty.

The Smallholder Optimists

This view, championed by Lipton (2004) and IFPRI (2002a and 2002b), takes a totally different tack. It states that the basic Green Revolution recipe is as valid today as it was in the past. With the right policies and support, smallholders will innovate and increase productivity, which will directly reduce rural poverty and stimulate wider economic growth and development through the transmission mechanisms described above.

Optimists argue that the classic links between agriculture and the wider economy that worked so well in Asia, can work elsewhere today and only the smallholder works at the scale required to have any meaningful impact upon poverty. Agriculture remains the most effective engine of growth and poverty reduction, and investing in the small-scale farm sector represents the most effective way of stimulating growth and reducing poverty.

Optimists acknowledge the changes in context emphasised by the pessimists, but question the extent to which they represent terminal damage to the future of small-scale farmers, for example:

- While global consolidation in food marketing has occurred, its impact on small-scale farmers, particularly in Africa, may be relatively insignificant as the value of staple food consumption in Africa exceeds that of export commodities by a factor of three (IFPRI, 2004).
- African farmers may face increasing difficulties operating in world markets, but abundant opportunities remain in domestic and regional markets for commodities such as maize, cassava and legumes, which are grown and consumed mainly by the poor. Furthermore, the local price of these goods is generally less affected by global price trends, because of high transport costs relative to value.
- Removal of state-provided input subsidies might have had less impact on the profitability of small-scale farming than would appear. A balanced assessment of the impact of liberalisation must take account of the frequent bias against output prices often implicit in state marketing systems, and the concealed taxation of farm sales through foreign exchange manipulation.
- While the non-farm economy is of greater importance to livelihoods of the rural poor than has perhaps been recognised in the past, evidence consistently shows that it is itself largely driven by the performance of the agricultural sector (Lanjouw and Lanjouw, 2001).
- For optimists, the bottom line is that 70% of the world's absolute poor are rural, and agriculture remains critical to their livelihoods, either directly through the on-farm economy or indirectly through agriculture's influence on the non-farm economy. Projections indicate that even by

2035 (well past the date for the MDGs) 50% of the poor are likely to be rural dwellers. Reducing poverty will only become feasible when the livelihoods of the rural poor are improved directly.

TWO VIEWS – TWO DIFFERENT SETS OF POLICY CONCLUSIONS

The main difference between pessimists and optimists lies in their views about the timing of economic transition and whether it is possible to by-pass the first stage in agricultural transformation, i.e. the stimulus provided by a major increase in agricultural productivity in staple food crops. Fundamentally.

- Has the impact of globalisation spread so far and fast that small-scale farmers in Africa and Asia can no longer compete in their local markets against imported grains?
- Have consumption patterns changed to such an extent, even amongst the poorest, that any increase in income will be spent largely on imported foodstuffs?
- Is international penetration of local markets, and market integration at a national and regional level, developing at such a speed that opportunities in agricultural production no longer exist for poorer communities? And if this is happening, what are the options for the poorest?

The policy implications of the two perspectives are quite starkly different. For *optimists* the issue is not one of whether smallholders can succeed, but how to make sure they do. This perspective emphasises the need for increased direct investment in agriculture and rural development, with support focusing on creating institutions that will encourage and support smallholder-led agricultural development. This includes the public provision of input and output marketing services, infrastructure development and democratic processes of land reform. Price incentives will need to be in place, but optimists argue that incentives can largely be created by removing developed country policies that have artificially depressed international commodity prices.

Particular attention will need to be given to improving the productivity of staple food crops that are not internationally traded, but consumed by the poor and traded locally. As research into these crops attracts little private sector attention, there will be a need for public funding with international assistance. In addition, research must take greater account of differing natural resource environments, in particular water management and soil degradation, and become more focused and more region specific. In the future, labour-intensive approaches may not be the most suitable as HIV/AIDS reduces labour availability in some African countries in particular.

For *pessimists*, the strategy is more or less the opposite. They emphasise the need to achieve the best outcome for the poor from current, rapidly changing trends. As far as agriculture is concerned, efforts should be focused on encouraging commercial production of non-staple cash crops, particularly those that result in robust links to the non-farm sector, as this will be the major provider of employment for the rural poor. Influencing international policy processes will be important, but primarily to ensure access to developed country markets for more processed and high quality products from developing countries. The rural poor will be best assisted by improving their access to health and education services to improve their human capital and through measures that increase their mobility so that they can move to take up opportunities in growth areas as they occur.

3

Productivity Growth in Indian Agriculture: The Role of Globalization and Economic Reform

The Indian agricultural sector has been undergoing economic reforms since the early 1990s in the move to liberalize the economy to benefit from globalization. This paper traces this process, analyses its effects on agricultural productivity and growth and discusses the problems and prospects for globalization to draw policy implications for the future of Indian agriculture.

India, which is one of the largest agricultural-based economies, remained closed until the early 1990s. By 1991, there was growing awareness that the inward-looking import substitution and overvalued exchange rate policy coupled with various domestic policies pursued during the past four decades, limited entrepreneurial decision making in many areas and resulted in a high cost domestic industrial structure that was out of line with world prices. Hence the new economic policy of 1991 stressed both external sector reforms in the exchange rate, trade and foreign investment policies, and internal reforms in areas such as industrial policy, price and distribution controls, and fiscal restructuring in the financial and public sectors. In addition, India's membership and commitment to World Trade Organization (WTO) in 1995 was a clear sign of India's intention to take advantage of globalization and face the challenge of accelerating its economic growth.

One measure of economic growth is given by productivity growth as it forms the basis for improvements in real incomes and welfare. The concept of productivity growth gained importance for sustaining output growth over the long run as input growth alone is insufficient to generate output growth because of diminishing returns to input use. This paper, which examines India's productivity growth in the agricultural sector in the context of globalization, has three main aims. First, it examines these possible links in the agricultural sector in general. Second, it discusses the problems and prospects for agricultural productivity growth of various Indian states. Third, the paper highlights the challenges of globalization and draws policy implications for the success of Indian agriculture.

OVERVIEW OF INDIA'S AGRICULTURAL ECONOMY

In the early 1950s, half of India's GDP came from the agricultural sector. By 1995, that contribution was halved again to about 25 per cent. As would be expected of virtually all countries in the process of development, India's agricultural sector's share has declined consistently over time as seen in the table below.

Table 1. Share of agricultural output in India's GDP

Year	*1950/51*	*1965*	*1976*	*1985*	*1991*	*1999*
Percentage share	52.2	43.6	37.4	32.8	28.3	24.4

Source: Estimated from various issues of Economic Survey, Government of India.

In the last five decades, the Government's objectives in agricultural policy and the instruments used to realize the objectives have changed from time to time, depending on both internal and external factors. Agricultural policies at the sectoral level can be further divided into supply side and demand side policies. The former include those relating to land reform and land use, development and diffusion of new technologies, public investment in irrigation and rural infrastructure and agricultural price supports. The demand side policies on the other hand, include state interventions in agricultural markets as well as operation of public distribution systems. Such policies also have macro effects in terms of their impact on government budgets. Macro level policies include policies to strengthen agricultural and non-agricultural sector linkages and industrial policies that affect input supplies to agriculture and the supply of agricultural materials.

During the pre-green revolution period, from independence to 1964-1965, the agricultural sector grew at annual average of 2.7 per cent. This period saw a major policy thrust towards land reform and the development of irrigation. With the green revolution period from the mid-1960s to 1991, the agricultural sector grew at 3.2 per cent during 1965-1966 to 1975-1976, and at 3.1 per cent during 1976-1977 to 1991-1992. Acharya (1998) explains that the policy package for this period was substantial and consisted of:

(a) introduction of high-yielding varieties of wheat and rice by strengthening agricultural research and extension services,

(b) measures to increase the supply of agricultural inputs such as chemical fertilizers and pesticides,

(c) expansion of major and minor irrigation facilities,

(d) announcement of minimum support prices for major crops, government procurement of cereals for building buffer stocks and to meet public distribution needs, and

(e) the provision of agricultural credit on a priority basis. This period also witnessed a number of market intervention measures by the central and state Governments.

The promotional measures relate to the development and regulation of primary markets in the nature of physical and institutional infrastructure at the first contact point for farmers to sell their surplus products.

Acharya (1998) also notes that the rate of growth of productivity per hectare of all crops taken together increased from 2.07 per cent in the decade ending 1985-1986 to 2.51 per cent per annum during the decade ending 1994-1995. Similar evidence of an increase in yields, a partial measure of productivity gains given by output per unit of land area is seen below for various crops.

Although productivity gains were sustained in the 1990s after the liberalization process began, the yield rates for most of the agricultural products in India are far below comparable rates in a number of other countries. This is seen in table 3. Except for sugarcane, tea, coffee and jute, India's yields are lower than the world average. It should be noted that India is ranked second both in area and output for sugarcane production and is the largest producer of tea and jute in the world. Although India is doing quite well in wheat production, the average yields in the Netherlands and Ireland are more than three times India's yield rates. In all other major crops, India's productivity performance seems to lag behind others.

Table 2. Yield for various crops (kg/ha)

	1950/51	*1960/61*	*1970/71*	*1980/81*	*1990/91*	*1995/96*	*1998/99*
Rice	668	1 013	1 123	1 336	1 740	1 855	1 905
Wheat	663	851	1 307	1 630	2 281	2 483	2 596
Coarse cereals	408	528	665	695	900	941	1 035
Pulses	441	539	524	473	578	552	661
Food grains	522	710	872	1 023	1 380	1 499	1 611
Oil seeds	481	507	579	532	771	851	948
Cotton	88	125	106	152	225	246	240
Sugarcane	33 422	45 549	48 322	57 844	65 395	68 369	69 288

Source: Agricultural Statistics At A Glance. Agriculture, CMIE, September 1999.

Why globalize?

Globalization in the context of agriculture can be best discussed in the context of three components – improvement of productive efficiency by ensuring the convergence of potential and realized output, increase in agricultural exports and value added activities using agricultural produce, and finally, improved access to domestic and international markets that are either tightly regulated or are overly protected.

These components are linked in various ways. For example, productive efficiency would enhance value added activities in agriculture through agro-processing and exports of agricultural and agro-based products. These activities in turn would increase income and employment in the industrial processing sector. Thus globalizing agriculture has the potential to transform subsistence agriculture to commercialized agriculture and to improve the living conditions of the rural community.

Table 3. Annual average yield based on 1995, 1996 and 1997 data

(kg/ha)

Region	*Crops*											
	Wheat	*Rice*	*Coarse Grains*	*Maize*	*Pulses*	*Ground-*	*Sugar-nuts*	*Tea cane*	*Coffee*	*Cows Milk*	*Tobacco Leaves*	*Jute*
World	2 565	3 757	2 678	4 050	4 050	1 318	63 266	1 158	538	2 062	1 616	1 757
Europe	4 764		4 311	5 698	5 698					3 557		
China	3 787	6 186	4 358	4 867	4 867	2 686	64 159	697		1 005	1 845	2 522
India	2 569	2 848	982	1 540	1 540	1 059	71 040	1 776	741		1 442	1 827
USA	2 508	6 590	6 590	7 690		2 787				7 543	2 275	
Canada	2 245											
Australia	1 880											
Thailand		2 252										
Indonesia		4 473						1 365	480			
Sri Lanka							67 718	1 385				
Brazil			2 482						582		1 473	
Kenya												
Bangladesh												1 580
UK	7 772							2 155				

Source: Food and Agricultural Organization Yearbook 1997.

However, economic reforms within India are necessary to pave the path to successful globalization. The stated objective of the new economic policy is to raise the economy's growth rate from the current 5.5 per cent achieved over 15 years to about 7 or 8 per cent per year.

Ahluwalia (1996) explains that this indirectly requires an improvement in agricultural growth from between 2 and 3 per cent in the past to about 4 per cent per year. Although initially, with respect to agriculture, there was no major policy reform package in the 1990s, it was however anticipated that the opening up of the agricultural sector to foreign trade, the move to a market determined exchange rate and reduction of protection for industry would, over time, benefit the agricultural sector.

Manmohan Singh (1995), the then Finance Minister, in his inaugural address at the 54th Annual Conference of the Indian Society of Agricultural Economics, brought to notice that a policy of heavy protection of the industrial sector operated to the disadvantage of the agricultural sector when industrial prices were raised relative to world prices and thus the profitability of investing in industry was raised relative to agriculture. This would lead to a shift of resources from agriculture to industry. A policy of heavy Industrial protection also led to an appreciation of the exchange rate. Ahluwalia (1996) noted that over-valuation of the exchange rate (before the Indian rupee was devalued by 18 per cent in two phases starting in July 1991) discouraged agricultural exports more than industrial exports because Indian industrial policy had sought to offset the constraints faced by industries via a system of export incentives for market support. Agricultural exports on the other hand were denied any such incentives as they did not use imported inputs.

Ahluwalia (1996) argued that in the past, the agricultural sector was negatively protected because of the above two reasons and the fact that farmers were denied access to the world markets due to trade barriers. Exports of plantation crops and a few commercial crops were free from export restriction but exports of essential commodities, particularly food products, were subject to bans, quotas and other restrictions. Interestingly, Kruger and others (1991) showed that while many developed countries continue to protect agriculture, developing countries do not do so. However, no formal attempt or theoretical framework has yet been used to assess the extent of negative protection in Indian agriculture. The implementation of economic reform in the Indian agricultural sector has been a gradual process. These include an 87 per cent cut in tariff on agricultural products, sustenance of high-yield crop varieties, removal of minimum export price on selected agricultural products, a lift on quantity restrictions on the export of some crops and various land reforms related to tenancy rights and land ceilings.

Productivity Gains from Globalization and Economic Reforms

In the wake of India's efforts towards globalization and economic reforms, the expected benefits of total factor productivity (TFP) growth can be represented using the production frontier. The production frontier traces out the maximum output obtainable from the use of inputs. In the figure below, F_1 and F_2 are the production possibility frontiers in time 1 and 2 respectively.

Opportunities from globalization and economic reforms can lead to:

a) shift from A to B due to technical efficiency

b) shift from B to C on existing frontier due to input growth

c) upward shift from C to D due to technological progress

Each of the above-mentioned shifts, which constitute various sources of TFP growth, can be linked with trade gains. The movement from A to B led by technical efficiency allows increases in output when inputs and technology are used to their fullest potential to obtain the greatest yield. Given that India has been involved in agricultural production for so long, there would be learning-by-doing gains that can help boost production given the expected increase in demand as India opens up. The increased production would enable a better utilization of inputs, especially that of advanced capital technology. The reduction in the tariff rate for agricultural products from 113 per cent in 1990-1991 to 26 per cent in 1997-1998 is also expected to motivate local producers into rethinking their production techniques and efficiently utilizing the inputs and technology to keep costs of production down in order to remain competitive. The optimum or efficient use of land and water resources would then allow agriculture to respond to the demand for other products such as horticulture and livestock which is expected to increase following a rising trend in the per capita incomes of both rural and urban groups.

Figure 1. Total factor productivity gains from globalization and economic reforms

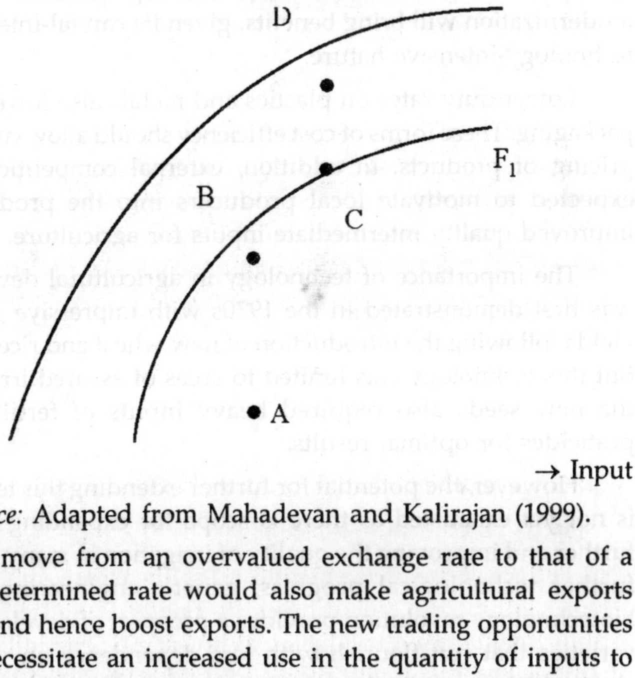

Source: Adapted from Mahadevan and Kalirajan (1999).

The move from an overvalued exchange rate to that of a market determined rate would also make agricultural exports cheaper and hence boost exports. The new trading opportunities would necessitate an increased use in the quantity of inputs to boost output and this allows for the movement from B to C along the existing production possibility frontier. Increased exports would bring about economies of scale and as Verdoon's law states, output growth would lead to productivity growth.

The scale of output under increased exports would justify the huge fixed costs underlying technologically advanced equipment and hence increase incentives to adopt high quality inputs. The use of such inputs would result in technological progress and this is represented by the shift from C to D. The reduction in tariff rates in industry from 1990-1991 to 1997-1998 range from 153 per cent to 25 per cent for consumer goods, 77 per cent to 18 per cent for intermediate goods and 97 per cent to 24 per cent for capital goods.

This means that farmers now have relatively cheaper access to imported new technology and better capital equipment as well as the option of adopting better farming techniques and this should lead to technological progress. In particular, the development of agro-processing as an instrument for agricultural and rural modernization will bring benefits, given its capital-intensive and technology-intensive nature.

Lower duty rates on plastics and metals also lower costs of packaging. These forms of cost efficiency should allow competitive pricing of products. In addition, external competition can be expected to motivate local producers into the production of improved quality intermediate inputs for agriculture.

The importance of technology in agricultural development was first demonstrated in the 1970s with impressive growth in yields following the introduction of new wheat and rice varieties. But this technology was limited to areas of assured irrigation as the new seeds also required heavy inputs of fertilizers and pesticides for optimal results.

However, the potential for further extending this technology is not yet exhausted as there is scope for expanding irrigation further and improving the quality of irrigation in many areas. For further technological progress, genetic engineering and the biotechnology revolution provides a prospect of developing new varieties that can flourish with less dependence on water and chemical inputs. Such reduced dependence upon chemical fertilizers and pesticides is also desirable because of environmental considerations, which are an increasing concern.

It must however be acknowledged that the link between trade liberalization and productivity growth is two-way as they both feed on each other. The above discussion has shown how productivity gains can be obtained from openness but to benefit from openness via increased demand for exports, agricultural products need to be priced competitively. In other words, productivity growth is necessary to lower the costs of production.

Agricultural Growth and Performance: An Economy-wide Analysis

Although India's economic reforms were initiated in June 1991, the process of liberalization was implemented gradually and thus it is difficult to assess the full impact of the liberalization measures. Nevertheless, an attempt is made to discuss what is observable in terms of agricultural growth.

One observation is that the expected increase in exports due to liberalization simply did not occur. India's share in world exports was 0.6 per cent in 1997; India has to aim for at least 4 per cent by 2005 in order to meet the growing import demands for capital goods, raw materials and crude oil as well as to meet her external financial commitments (Kalirajan and others 2001). For the last decade or so, India's share in world exports of agriculture has been between 2 per cent and 3 per cent. Furthermore, as table 3 shows, India is not as competitive as the other countries and calculations show that India's crop yields have increased at a slower rate over the 1990s.

In addition, the agricultural sector's output growth decreased to 2.9 per cent during 1992-1993 to 1998-1999. Kalirajan and others (2001) explain that two important reasons for the slowdown are that there was no major breakthrough in developing new high-yielding varieties during the 1990s and there was a decline in the environmental quality of land which reduced the marginal productivity of the modern inputs. What could this mean in terms of the effectiveness of the policies of reduced protection to industry, a market determined exchange rate and the opening of the agricultural sector to foreign trade?

First, although the reduction to protection of industry is substantial, there is reason to believe that the reduction was not necessarily sufficient to benefit the agricultural sector whose tariffs were also drastically reduced. Hence, the expected shift in resources to agriculture did not occur. Second, is the apparent ineffectiveness of the market determined exchange rate in boosting exports. This is however not surprising as the exchange rate may not be a key factor determining agricultural export demand for India. In general, unlike manufacturing industries, agriculture did not benefit much from these two policies because the share of imported inputs in the value of agricultural production is small. It is likely that a change in the mindset and attitude of farmers has yet to take place and there are delays or hesitation in embracing India's openness.

Third, in opening up the agricultural sector to foreign trade, India has taken major steps towards trade liberalization since 1991, partly on its own initiative and partly from its commitments to WTO. Kalirajan and others (2001) provide a detailed review of these reform procedures. But why have the benefits from trade liberalization been slow to come?

One reason is that prospects for growth in agricultural exports depend partly on domestic policies and partly on the removal of protectionist policies pursued by developed countries such as Japan and members of the European Union (EU). An OECD report (1998) estimated that the producer equivalent subsidy in the OECD countries increased by US$ 9.3 billion from 1988 to 1993 and this subsidy as a percentage of the value of production in 1997 was 9 per cent in Australia, 20 per cent in Canada, 47 per cent in EU and 70 per cent in Japan. These protectionist practices do not seem likely to come to an early end. An UNCTAD report (1999) noted that 29 member countries of the OECD spent an average of US$ 350 billion a year in agricultural support between 1996-98. Schumacher (2000) further reports that the EU provides product-specific trade distorting domestic support to at least 50 different agricultural products. The implication of these reports is that food exports from India may not show a large increase given the international environment and the still-existing restrictions on exports in the major importing markets based on the self-sufficiency argument and food security. Other macroeconomic factors, such as the recession in developed countries in 1996-98 as well as the 1997 South-East Asian financial crisis, have clouded the possibilities of increasing Indian exports.

Another problem faced by Indian agricultural exporters is the protectionist measures in the form of non-trade barriers that developed countries use to restrict market access. This is by tightening requirements of quality, testing and labeling, and anti-dumping and countervailing measures. For example, in May 1997, the EU banned marine products from India citing unhygienic processing conditions. The extra costs of meeting the standards required in export markets as well as costs associated with changes in the production mix and transactions associated with exports may well be discouraging Indian exporters.

One existing problem of India's agricultural protection is the use of input subsidies. The general argument favouring this has been that it is necessary to encourage the use of particular inputs for production for various benefits. For India, Gulati and Sharma (1995) show that the input subsidy in per cent of GDP increased from 2.13 in the triennium ending 1982-1983 to 2.73 in the triennium ending 1992-1993.

But the benefits of these subsidies have accrued to only certain classes of farmers in some regions cultivating irrigated crops. Furthermore, highly subsidized prices of inputs such as irrigation water and electricity for pump sets have encouraged cultivation of water-intensive crops, over-use of water, ground water depletion/salinity and water logging in many areas. Subsidy for nitrogen fertilizer on the other hand has resulted in nitrogen phosphorous potassium imbalance and acted as a disincentive for use of the environmentally friendly organic manure. As a result, the linkage between food crops and non-food crops, which include fodder, has been reduced.

These adverse consequences are a drain on the fiscal burden of central and state Governments. Thus, if not properly monitored, input subsidies can be counterproductive and, in this context, protection to lower costs of production should be done selectively in the course of liberalization.

In fact, Agenda 21 of the United Nations Conference on Environment and Development in 1992 stressed that there is a need for integration of environmental considerations in the pricing of natural and other resources in such a way that prices reflect social costs.

Such a pricing policy will not only lead to a more efficient use of scarce resources but also result in subsidy reductions and improvements in environmental quality. The money saved from the reduction of subsidies can be spent in the development of rural infrastructures, agricultural research, farmers' education and other forms of support for agriculture.

Agricultural Growth and Performance

While the above analysis has provided a general view of the impact of economic reforms, this section examines agricultural growth and performance in the states of Bihar, Karnataka, Tamil Nadu and Punjab with their attendant policy implications. The table below shows the yield for various crops in these four states.

Table 4. Yield of major crops (kg/ha)

Year	*Rice*	*Wheat*	*Coarse cereals*	*Pulses*	*Food grains*	*Oil seeds*	*Cotton*	*Sugar-cane*
				Bihar				
1960/61	867	679	730	515	749	398	-	33 879
1970/71	788	957	885	600	795	458	-	38 353
1980/81	1 015	1 314	828	609	989	423	-	31 412
1990/91	1 220	1 810	1 114	791	1 300	617	-	52 490
1995/96	1 370	2 020	1 566	610	1 450	620	-	45 510
				Karnataka				
1960/61	1 292	-	441	270	518	407	389	70 149
1970/71	1 709	-	696	354	774	626	499	78 689
1980/81	2 029	-	793	319	873	520	590	78 232
1990/91	2 070	-	780	333	888	525	1 074	76 287
1995/96	2 380	-	1 294	470	1 290	680	980	79 560
				Punjab				
1960/61	1 009	1 244	904	785	1 032	654	270	36 541
1970/71	1 764	2 237	1 411	744	1 860	790	350	41 171
1980/81	2 733	2 730	1 548	589	2 561	786	308	55 211
1990/91	3 229	3 715	1 907	755	3 390	958	463	59 410
1995/96	3 130	3 880	1 995	820	3 840	1 200	440	65 300
				Tamil Nadu				
1960/61	1 413	-	787	265	1 058	900	167	80 000
1970/71	1 900	-	784	271	1 340	919	196	77 367
1980/81	1 861	-	841	324	1 340	846	201	100 820
1990/91	3 116	-	1 106	425	1 910	1 081	290	113 920
1995/96	3 390	-	1 154	370	2 140	1 470	350	110 010

Source: Statistical Abstract, Government of India.

Tamil Nadu: Season and Crop Reports.

Table 4 shows that the yields for various crops in these states differ greatly. While Tamil Nadu had the highest yield in rice, oil seeds and sugarcane, Punjab enjoyed the highest yields in wheat, coarse cereals, pulses and food grains. Karnataka on the other hand is seen to do well in cotton and Bihar performed quite well in pulses and coarse cereals. Further analysis and findings by Kalirajan and others (2001) show that Punjab had made remarkable achievements on the agricultural front while Bihar had remained stagnant in the last two decades, with Karnataka and Tamil Nadu showing moderate achievement. Clearly, differences in physical endowments, climatic conditions and institutional characteristics are some of the reasons for the varying productivity performance. Thus, having across the board economic reforms is likely to work less effectively than state-specific policy measures that enable each state's agricultural yields to reach their full potential. The comparative advantage of each state's agricultural production should be determined and with inter-state restrictions removed, total agricultural output would see a very significant increase.

For example, Karnataka with less favourable soil and water resources should be given incentives to concentrate on agro-processed products and corporate agriculture in horticulture, floriculture and animal husbandry, or to undertake watershed development to help with dry land agriculture. Many studies have indicated that with watershed areas, productivity growth has been mainly due to seed and fertilizer use. Thus, this state has to be given input subsidies for high yielding seed varieties but at the same time, the farmers need to be educated on the over use of chemical fertilizers.

With Bihar, agricultural performance is problematic on many fronts. First, although demographic pressure has increased and agricultural technology has improved, most of the uncultivated land is concentrated in southern Bihar, where irrigation facilities have not kept pace and the soil is of poor quality. Given the physiography of southern Bihar, wells are also unsuitable and thus the dominant mode of irrigation has been through tanks whose expansion and maintenance has been neglected. Second, the infrastructural facilities of Bihar have been lagging as seen by the infrastructure development index in table 5. Due to infrastructural bottlenecks, availability of modern goods and services has not increased or their supply remains costly or unreliable.

Table 5. Infrastructure development index

(All India = 100)

	Bihar	*Karnataka*	*Punjab*	*Tamil Nadu*
1980/81	83.5	94.8	207.3	158.6
1984/85	84.8	97.9	204.6	148.7
1989/90	83.1	95.2	195.8	147.4
1992/93	83.4	96.1	191.6	143.3
1993/94	81.1	96.9	191.4	144.0

Source: Profiles of States, CMIE, March 1997.

Third, agriculture in Bihar is dominated by small and marginal farmers and the prevalence of mass poverty is largely related to the backwardness of agriculture. Fourth and importantly, the state agricultural policies in Bihar are in dire need of review. The semi-feudal production condition still exists in rural areas and the ineffective protection of tenancy rights has hindered agricultural growth. The slow pace of land consolidation reflects inadequate financial outlays and a shortage of manpower. Kalirajan and others (2001) note that marketing and extension services in Bihar are also rather weak compared to the other states.

Punjab on the other hand, was one of the few states which enjoyed the success of land reforms and the high priority of investment in rural infrastructure as seen in table 5. Also, the irrigation base of the small and medium sized farms was comparable to that of large farms. In addition, the Punjab Agricultural University at Ludhiana contributed to the development of new seed varieties.

However, there are clear signs of a decline in crop yields since the 1990s and this has been associated with the increasing use of fertilizers and excessive water use which have increased the unit cost of production as a result of declining soil quality. Hence, care is needed when providing further input subsidies in fertilizer and water use. Another related fact is the steep increase in wages in

Punjab and in the absence of productivity increases, the cost increase has affected the profitability of farmers.

With Tamil Nadu, the main crop has been rice as this state is blessed with two monsoons. But from 1992-1997, there has been a steady decline in the areas irrigated by canals and an increase in well-irrigated areas while the use of tanks remains an unreliable source of irrigation. However, major improvements in about 10 rice varieties released in the early 1990s can be expected to improve productivity growth in rice production although pests and diseases as well as imbalance in the use of fertilizers are major constraints. Thus Tamil Nadu could do with subsidies of pesticides and farmers should be educated on the more effective use of fertilizers to obtain high yields. Interestingly, the cropping pattern of late has shown increasing substitution of food crops by commercial crops but there is concern that the benefits will reach farmers only with the development of adequate infrastructure such as roads and markets. Table 5 shows, however, that Tamil Nadu has a higher index than the all India average of infrastructure.

Challenges of globalization

It is important to realize that globalization poses many challenges to a developing country like India, which had relied on a state directed and regulated policy regime for more than four decades. In moving to a more open, market-based economy there are many transitional problems that the country has to manage. The Government must play a pro-active role in facilitating the globalization process so that the opportunity sets for the economic agents are widened and the adverse effects of globalization are minimized. The Indian Government must also prepare the necessary information base and develop its capacity to articulate India's concerns and policy trade-offs in the international forums for multilateral trade and environmental negotiations.

In addition, the Government should embark on an extensive programme to educate farmers on the need to meet the standards required in the export markets. In fact, India needs to seek technical assistance in creating the capacity for meeting such standards and to consider watershed developments for environmental considerations. Equally important is the need to disseminate information about possible export markets to farmers, so that market access is achieved at minimum cost. Given the requisite information about markets and profitability, the likelihood of farmers investing in post-harvest and processing technologies and storage and efficient transportation arrangements as well as developing supporting infrastructure is very high.

Although the brave and bold move by India to reduce the tariff rate for agricultural products from 113 per cent in 1990-1991 to 26 per cent in 1997-1998 deserves to be applauded, the question of whether India is ready to compete in world markets remains to be seen. The infant industry argument may still hold for India to shield itself from external competition but one can easily question the length of time that is required to that end. Also, a delay in opening up to foreign trade has the danger that local producers may become too complacent and never be ready for competition.

As India opens up externally, it is also expected to face vulnerability in the wider international price fluctuations and thus Acharya (1998) claims that a minimum price support scheme is important. These prices can also act as a signal to adopt modern inputs and invest in yield-raising infrastructure for increasing production. For instance, keeping basic staple food grains at reasonable prices would induce farmers to switch over to high value crops. However, during the 1990s, Kalirajan and others (2001) shows that procurement prices especially for rice and wheat have been increasing faster than the general price level. Such high prices along with guaranteed purchases by the Food Corporation of India have pushed up market prices. These higher prices are partly responsible for the large buffer stocks with the Food Corporation. If this trend continues, India's comparative advantage will be eroded.

With openness and high price instability, unstable export revenue can also be expected. One way of reducing such risk is for India to diversify her agricultural exports. For example, since 1990, even in commodities such as tea, coffee, cocoa and spices, where India is supposed to have a comparative advantage (Chadha, 1999) international prices have been unstable. Besides increasing the type of exports to obtain more export revenue, India should also seriously consider exporting more value added agricultural products through agro-processing such as processed vegetables, fruits, fish and meat products given that export or even local demand for basic agricultural products would decline as incomes rise. The move to higher value added activities within the agricultural sector also spells greater opportunities for industrialization and vice versa as borne by Kalirajan and Shand's (1997) findings of a bi-directional relationship between agriculture and industry for most Indian states. On the other hand, Sivakumar and others (1999) establish empirical evidence of high forward linkages of agriculture due to the presence of agro-industries while Satyasai and Viswanathan (1999) show the significance of the spillover effects to the industrial sector via the intensive use of purchased inputs in the agricultural sector.

The lack or slow pace of internal or domestic liberalization is also seen to hinder the possible gains from external or trade liberalization. For example, although central zoning restrictions have been abolished, state government restrictions on inter-state and even inter-district restrictions on marketing and movement of goods still exist in many cases. This interferes with the benefits from crop specialization and economies of scale arising from comparative advantage. The land market is another example of distortion whereby land ceilings exist preventing the operation of large-sized farms. This has led to the emergence of a large number of small economically unviable land holdings. The easy leasing of land should be permitted with assurance of resumption. Yet another problem lies with the insufficiency of credit to agriculture. From 1995-1996, the Rural Infrastructure Development Fund was set up to allocate funds for the completion of projects and the government has committed itself to strengthening the cooperative credit structure through substantial refinancing and restructuring of the Regional Rural Banks. However, as mentioned earlier, due to varying institutional factors in the Indian states, these domestic reforms can be expected to yield quite different results.

CONCLUSION

Although India missed the opportunity to open up two decades ago, its attempts to do so now must be regarded as better late than never. Others such as Desai (1999) observe that, "the logic of the global economy as well as India's interests dictate that India become proactive in its liberalization policies. India must liberalize not because it has no choice but because it is the best choice".

His lament that India has adopted a 'victim mentality' when it really needs to adopt a 'winner mentality' has become less of a concern as over time, India has shown commitment to stay on the bandwagon of globalization. Having realized that globalization is a necessary but not a sufficient condition for high growth production, India has undertaken economic reforms, both internal and external. However, it must be ensured that these reforms are synchronized so that the pace of both reforms is set right in order to work hand in hand to promote agricultural productivity growth.

Thus, training the farmers and educating them appropriately to change their mindset and reorienting them to take up new activities or adopt foreign technology is of utmost importance. In this context, it is necessary to involve non-governmental organizations in training and mobilizing the rural poor to face the challenge of liberalization.

Also, with domestic economic reforms, more care needs to be exercised to draw up state-specific liberalization measures to maximize their benefits. Lastly, in the implementation of these reforms for successful globalization, one crucial element, not entirely within control is the need for good governance and stability in the political and economic environment. Political leaders who are the ultimate decision makers in these matters need to examine their own role dispassionately.

It is quite apparent that at this relatively early stage, there is little observable evidence of gains to India's agricultural performance after opening up. However, there could easily be benefits that have not yet surfaced, or are yet to be identified and perhaps too difficult or intangible to measure.

Whatever the case, it is highly likely that it is too soon to assess the full impact of globalization and economic reforms. Furthermore, the process of liberalization has been gradual and remains incomplete. For example, the complete removal of quantitative restrictions after March 2001 will have provided an opportunity for Indian farmers to tap world markets and, if they are successful, results should start to become evident soon. Export promotion via the development of export and trading houses as well as effective liberalizing export promotion zone schemes for agriculture are fairly recent measures and only time will tell as to how effective these measures are. Other possibilities such as agro-industry parks for promoting exports are also in the pipeline.

In conclusion, India has successfully set sail on the waters of globalization and economic reforms and even in the wake of economic and political instability, she has to carefully steer her course in order to reap the benefits of increased productivity growth in the agricultural sector.

4

Accelerating Agricultural Growth

Poverty in India is predominantly a rural phenomenon. About 70% of the population, and about 85% of the poor, live in rural areas and most depend on agriculture. Agriculture provides livelihood to 60 percent of the rural people and remains vital for food security.

KEY ISSUES AND CONSTRAINTS

Deceleration in agricultural growth. In the last decade (1995/96-2004/05) the agricultural GDP growth rate slowed to less than 2% per year, compared to about 3.5% per annum in the preceding decade. In the poorest states, such as Madhya Pradesh, Orissa, and Rajasthan, growth in the last decade was below 1% per year. The stagnation of agriculture, the high proportion of poor dependent on it, and the widening rural-urban income gaps are major concerns of the Government of India (GOI).

Low Agricultural Productivity. With fixed land and water availability, higher agricultural growth can be achieved only by increasing productivity per unit of these resources through effective use of improved technology. Per hectare yields of major crops (foodgrains, oilseeds, other cash crops) in India are lower than many other major producing countries and well below those obtained in crop trials within India.

Moreover, crop yields in many of the poorest states (Bihar, Orissa, Rajasthan, North-east states) are far below those of richer states (Punjab, Haryana, Tamil Nadu). The virtual collapse of the agricultural extension system in most states limits farmer access to improved technologies and practices. Weaknesses in the agricultural research system, such as an irrigated agriculture bias, weak prioritization and proliferation of programs, top-down supply driven programs, weak cost effectiveness, undermine the effectiveness of the public agricultural research system. Natural resources degradation.

Sustainability of the land and water resource base is at risk with increasing soil degradation and overexploitation of groundwater in many areas.

In addition to erosion, salinity and alkalinity, soils are also losing carbon and micronutrients due to unbalanced fertilizer use. Nearly 30% of the blocks in the country are presently classified as semi-critical, critical or overexploited as groundwater use exceeds the rate of groundwater recharge.

Composition of public expenditures. Public expenditures on agricultural subsidies including for power, irrigation and fertilizers, have crowded out productivity-enhancing investments, such as for irrigation development, agricultural research and extension, rural roads and electrification. According to the Planning Commission, budgetary subsidies in agriculture increased from around 3% of agriculture GDP around the late-1970s to about 7% in the early 2000s. During the same period public investment in agriculture declined from 3.4% of agriculture GDP to 1.9%.

GOVERNMENT OF INDIA'S RESPONSE

The Government of India's (GOI) goal is to raise agriculture's growth rate to 4% per year in the 11th Plan. To achieve this, the Plan aims to:

- Double the growth rate of irrigated areas and improve water management in rainfed areas
- Reclaim degraded land
- Bridge the knowledge gap through effective extension
- Foster diversification to higher value agricultural products
- Promote animal husbandry and fisheries
- Facilitate access to credit at affordable rates
- Improve the incentive structure for markets.

These issues are discussed in more detail below.

Water Resource Management and Irrigation. Water is a critical input for achieving higher agricultural growth and ensuring greater food security. Presently about 40% of cultivated area in India is irrigated. GOI is putting emphasis on more effective use of existing irrigation potential, expansion of irrigated area, and better water management in rainfed areas.

The GOI Accelerated Irrigation Benefits Program aims to assist states to finish the construction of uncompleted major and medium irrigation schemes. Many states, however, lack the policy, regulatory, and institutional framework for efficient, sustainable, and equitable allocation and use of water, or for capturing the environmental costs of inefficient use.

Many states often do not allocate sufficient public funds for operations and maintenance of canals, which lead to the rapid deterioration of irrigation canals and reduce the availability of water to farmers. Limited cost recovery also limits funds for operations and maintenance and undercuts farmer incentives to use water more efficiently, leading to waterlogging and salinity problems in some areas. Some states have adopted participatory irrigation management on a wide scale to improve the management and sustainability of surface irrigation systems.

Free power to farmers by some states or highly subsidized power encourages the excessive use of ground water. This has led to the increased proportion of over-exploited areas in the country and large fiscal costs to state governments. The political economy of power sector reform however impedes progress in this area.

Rainfed areas, where large numbers of poor live, cover 60% of cultivated area in India. Watershed management, rainwater harvesting and ground water recharge can help augment water availability in rainfed areas.

The GOI recently established the National Rainfed Areas Authority to develop an action plan for rainfed areas. In moving forward, some challenges that would need to be addressed include: (i) multiplicity of programs and agencies (MoA, MoRD, MoEF, donors) and lack of consistency among programs within the state (scope, community participation, contributions); (ii) weak coordination among various agencies; and (iii) weak local institutions to ensure effective implementation.

Bridging the Knowledge Gap. The GOI is providing additional resources for the National Strategic Research Fund. However, it would be critical to improve the governance and implementation structure of the Indian Council of Agricultural Research system to improve its research effectiveness.

The Ministry of Agriculture (MoA) is seeking to improve extension effectiveness by scaling up the Agricultural Technology Management Agency (ATMA) approach which builds on a bottom-up approach to extension planning and implementation.

It would be important to build on the factors of success derived from the ATMA pilots which include decentralized decision making and funds flow, priority to institutional capacity building and training of extension staff on the new bottom-up approach to extension planning and implementation and ensuring convergence of funds from the various departments at the district level.

The GOI is also encouraging states to develop district plans for agricultural development based on agro-ecological potential. These district plans will be the basis for the formulation of the state agricultural development plans which would be supported under the proposed additional central assistance of Rs. 250 billion announced by the Prime Minister in May 2007. Ensuring adequate stakeholder consultation and community participation in the development of these plans would strongly influence the success of this program.

Diversification of Production and Income Sources. Rising incomes are driving the increased demand for higher-value fresh and processed agricultural products in domestic markets and globally.

This opens new markets and income opportunities for farmers. Many of these commodities, however, are highly perishable and inadequate rural infrastructure (markets, rural roads, electrification) and services (market information, risk management) increase losses and inhibit the more rapid expansion of high value production. The GOI recently launched the National Horticulture Mission —

the single largest agriculture plan scheme — to promote increased horticulture production.

The Bharat Nirman Program, which supports the expansion of irrigation, rural roads, electrification and telecommunication) will help reduce losses and marketing cost and improve the competitiveness of local products.

The MoA is also encouraging states to amend their Agricultural Produce Marketing Committee (APMC) Acts to liberalize the marketing of agriculture produce at the state level and improve farmer access to markets. So far 11 states have adopted the amendments. In addition, MoA is also promoting public-private partnerships to develop markets and is providing investment grants to encourage investments in agro-processing and marketing.

Continuing trade and domestic regulatory reforms are encouraging increased investments in rural areas. These include the removal of most domestic marketing controls—storage, movement, credit, wholesale marketing, and food processing. However, reforms of the institutional and policy framework for land are needed.

Many states have computerized their land records/registration, but reforms of the institutional and data structures are essential to reduce the cost of land transfers and enhance tenure security in rural and urban areas (for example, eliminate overlapping property claims and facilitate access to credit). The bans or restrictions on land leasing limits access to land by poor and landless rural households and drives tenancy underground. They also limit the productivity of land use.

Diversification of incomes through rural non-farm activities is an important element of the overall strategy to address rural poverty. More rapid growth of the rural non-farm sector is, however, constrained by government interventions in labor, land, and credit markets and inadequate and poor quality infrastructure and services. To improve access to credit in rural areas the GOI is supporting reforms of the credit cooperative system.

Ensuring Food Security. While pursuing crop diversification, an emerging concern is the stagnation in foodgrain production which is raising food security concerns. The 11th Plan approach paper notes that to achieve the goal of a 4% overall agricultural growth rate per year, foodgrain yields per hectare need to increase by at least 2% to 3% per year to compensate for the possible shift in cultivated area from foodgrains to high value crops. To enable this, a concerted effort is needed to raise the relatively low foodgrain yield levels in Eastern and Central India in particular.

Improving Public Expenditures. In recent years the GOI and state government have exerted greater effort to cap the amount of the various subsidies, while increasing expenditures on more productivity enhancing investments. The latter includes several centrally sponsored programs, such as the National Horticulture Mission, Bharat Nirman rural infrastructure program, and the National Rural Employment Guarantee Scheme. Participation of the poorest states in these programs is hampered, however, by lack of capacity to develop appropriate plans and implement them effectively.

5

Objectives, Strategy and Perspective of Development

INTRODUCTION

The Ninth Five Year Plan, launched in the 50th year of India's Independence, will take the country into the new millennium. Much has happened in the fifty years since independence. The people of India have conclusively demonstrated their ability to forge a nation united despite its diversity, and their commitment to pursue development within the framework of a functioning, vibrant and highly pluralistic democracy.

In this process democratic institutions have put down firm roots and flourished and development has also taken place on a wide front. As the millennium draws to a close, the time has come to redouble our efforts at development, especially in the social and economic spheres, so that the country will realise its full economic potential and the poorest and the weakest will be able to shape their destiny in an unfettered manner. This will require not only higher rates of growth of output and employment, but also a special emphasis on all-round human development, with stress on social sectors and a thrust on eradication of poverty.

The Approach Paper to the Ninth Five Year Plan, adopted by the National Development Council, had accorded priority to agriculture and rural development with a view to generating adequate productive employment and eradication of poverty; accelerating the growth rate of the economy with stable prices; ensuring food and nutritional security for all, particularly the vulnerable sections of society; providing the basic minimum services of safe drinking water, primary health care facilities, universal primary education, shelter, and connectivity to all in a time bound manner; containing the growth rate of population; ensuring environmental sustainability of the development process through social mobilization and participation of people at all levels; empowerment of women and socially disadvantaged groups such as Scheduled Caste, Scheduled Tribes and Other Backward Classes and Minorities as agents of socio-economic change and development; promoting and developing people's participatory bodies like Panchayati Raj institutions, co-operatives and self-help groups; and strengthening efforts to build self-reliance. These very priorities constitute the objectives of the Ninth Plan.

Some specific areas from within the broad objectives of the Plan as laid down by the NDC have been selected for special focus. For these areas, Special Action Plans (SAPs) have been evolved in order to provide actionable, time-bound targets with adequate resources. Broadly, the SAPs cover specific aspects of social and physical infrastructure, agriculture, information technology and water policy.

The Ninth Plan is based on a careful stock taking of the strength of our past development strategy as well as its weakness, and seeks to provide appropriate direction and balance to the socio-economic development of the country. The principal task of the Ninth Plan will be to usher in a new era of growth with social justice and participation in which not only the Governments at the Centre and the States, but the people at large, particularly the poor, can become effective instruments of a participatory planning process. In such a process, the participation of public and private sectors and all tiers of government will be vital for ensuring growth with justice and equity.

OBJECTIVES OF THE NINTH PLAN

The Ninth Plan recognises the integral link between rapid economic growth and the quality of life of the mass of the people. It also recognises the need to combine high growth policies with the pursuit of our ultimate objective of improving policies which are pro-poor and are aimed at the correction of historical inequalities. Thus the focus of the Ninth Plan can be described as: "Growth with Social Justice and Equity".

The specific objectives of the Ninth Plan as approved by the National Development Council are as follows :

i. Priority to agriculture and rural development with a view to generating adequate productive employment and eradication of poverty;

ii. Accelerating the growth rate of the economy with stable prices;

iii. Ensuring food and nutritional security for all, particularly the vulnerable sections of society;

iv. Providing the basic minimum services of safe drinking water, primary health care facilities, universal primary education, shelter, and connectivity to all in a time bound manner;

v. Containing the growth rate of population;

vi. Ensuring environmental sustainability of the development process through social mobilisation and participation of people at all levels;

vii. Empowerment of women and socially disadvantaged groups such as Scheduled Castes, Scheduled Tribes and Other Backward Classes and Minorities as agents of socio-economic change and development;

viii. Promoting and developing people's participatory institutions like Panchayati Raj institutions, cooperatives and self-help groups;

ix. Strengthening efforts to build self-reliance.

The development strategy of the Ninth Plan has been tailored to the requirements of these objectives.

THE DEVELOPMENT STRATEGY

India's development strategy, like that of most other developing countries, has evolved over successive Plan periods, reflecting the growing strength of our economy, structural transformations taking place in the domestic economy and also developments in the world economy. In the early stages of development planning, government was viewed as the principal actor in development exercising strict control over private investment, ensuring a dominant role for the public sector in all important industries.

Trade policy tended to be inward oriented focusing on industrial development through import substitution which was encouraged through a tight control over imports and maintenance of high tariffs. The limitations of this strategy became evident by the end of the 1970s and early 1980s when it became clear that these policies reduced efficiency and competitiveness and growth was much lower than targeted. While government was over-active in industry, it was under-active in many areas, especially relating to social development and this was reflected in a very slow pace of improvement in critical social indicators.

Some efforts were made to reform the system in the second half of the 1980s to address the shortcomings in our development strategy. However it was not until 1991 that a wide ranging programme of economic reforms aimed at decontrolling and debureaucratising the economy was initiated. These reforms have been pursued by successive governments since 1991 and enjoy a broad base of support. They have also yielded good results thus far. There is no doubt that the Indian economy has responded well to the change in policy direction and the growth rate increased from 5.8% in the Seventh Plan (1985 to 1990) to 6.8% in the Eighth Plan. And yet there are many dimensions in which performance has lagged behind expectations.

Faster growth has not reduced poverty as much as it should have, nor has it created the number of high quality jobs we need to satisfy the aspirations of our increasingly educated youths. Growth has not been as regionally balanced as it should have been. The deficiencies in social development indicators have also continued and our low level of social development is today a major constraint on reaching a growth rate of 8 per cent, which should be our medium term target. The Ninth Plan is being implemented at a time when we are about to enter the new millennium. The international environment is also full of uncertainties in the aftermath of the East Asian crisis. It is appropriate at this critical period to articulate a development strategy which reaffirms and builds upon what has worked well, initiates corrective steps where needed, and takes new initiatives to meet the new challenges which face the economy in the years ahead.

There should be no doubt that the process of reforms which has been underway for some years, and which has yielded many good results, must continue and be strengthened. The reforms involve a major re-orientation of the role of the State. Instead of being a pervasive controller of private sector activity, and also a direct producer in many areas through majority owned public sector enterprises, the State must play a different role in future. One of the strengths of our economy is that we have a strong and vibrant private sector, including large, middle sized and small enterprises. Our agriculture has always been based on individual farmers with a predominance of small and marginal farmers. Our development strategy must be oriented to enabling our broad based and varied private sector to reach its full potential for raising production, creating jobs and raising income levels in society. A vigorous private sector, operating under the discipline of competition and free markets, will encourage efficient use of scarce resources and ensure rapid growth at least cost. Our policies must therefore create an environment which encourages this outcome.

Encouragement of the private sector and reliance upon competition and market forces does not mean that the State has no role to play or even that its role in development may be reduced. What is involved is not so much a reduction in the role of the State as a re-orientation. The State must withdraw from the role of being a controller and licensor of private enterprise in areas where market competition and an efficient financial sector will ensure appropriate decisions on investment and technology. However, there are many areas where the State must actually increase its involvement. The most important of these is in the area of social development especially in rural areas. Despite progress in the other areas over the past decades, we are woefully lacking in providing basic services such as health care, education, safe drinking water etc. to the majority of our population especially in rural areas. Extension of these services to the mass of our population is an urgent priority, not only because it is a desirable social end in itself, but also because it is a precondition for achieving rates of growth of 7-8% per annum. Provision of these services to the mass of our people can only be achieved through a much larger role to be played by the State. The provision of economic infrastructure such as power, roads, ports, railways, telecommunications, municipal services etc. is another area where the State will need to play a continuing role. Some of these services can be provided by the private sector and it is essential to tap this potential as much as possible so that the deficiencies in these areas can be met as quickly as possible. However, even with the best possible effort to attract private investment in infrastructure the State will continue to have to provide a large part of the investment needed for economic infrastructure. Public investment in infrastructure must continue to have high priority especially in those areas of infrastructure where the private sector's continuation is likely to be limited. There are other areas where the State must continue to play a major role and these are examined in detail in the next section.

Agriculture provides the livelihood of the largest number of our people and robust growth in this sector is the best guarantee of achieving a broad based growth of income levels and employment especially in rural areas. Our development strategy must focus special attention on this sector which requires a unique combination of private effort and public support. Agriculture has not benefited as much as it should have from policies of economic liberalisation because agriculture continues to suffer from too many restrictions and impediments which prevent farmers from marketing their produce at attractive prices. The phased reduction in the high levels of protection given to industry, and the opening up of export market for agricultural products, will help to shift the terms of trade in favour of agriculture and this should help to raise rural incomes. But broad based agricultural development also requires substantial investments in economic infrastructure especially irrigation, rural roads and creation of organised markets. These are all areas where the State must play a lead role. Agricultural research is another area where the role of the State will continue to be important.

The process of liberalising Central government control over the industrial sector has advanced considerably. Industrial licensing has been eliminated for all but a handful of industries and the list of industries reserved for the public sector has also been sharply reduced. The remaining controls at the Central Government level need to be reviewed for further liberalisation. Indian industry must be unshackled from unnecessary bureaucratic and governmental interference. However the major effort in future has to be to extend liberalisation to the level of State Governments. Indian industry still suffers from a plethora of controls and regulations relating to matters in the purview of State Government which cumulatively impose a heavy burden of delay and even harassment of our entrepreneurs. A thorough revamp of these controls and the procedures involved at the State Government level would help to create a climate in which Indian industry can flourish.

The role of the public sector as a direct producer of goods and services needs to be reviewed in the context of experience gained and emerging realities. The Eighth Plan saw the start of a process of disinvestment of government equity from public sector enterprises and the resources so generated were used to supplement available budgetary resources to finance development expenditure. Initially, disinvestment was undertaken subject to government equity remaining at 51%. There is no particular reason why government should retain a majority equity stake in public sector enterprises except for those that are in the strategic areas where national security is involved. For all other enterprises the government has announced that as a general rule it will dilute its holdings to 26 per cent as early as possible. The manner in which this is to be done will be determined on the basis of recommendations of the Disinvestment Commission. The Commission has made recommendations regarding several public sector enterprises including recommendations to induct a strategic partner who would exercise full or joint management control, in some cases. These recommendations need to be implemented expeditiously. The resources realised through sale of government equity in PSUs could be used to finance much needed investments in the social sectors especially health and education. At a time when resources are scarce, there is merit in unlocking resources embedded in public sector assets created through earlier investments in these enterprises and using these resources to invest in the social sectors, where we are seriously deficient. Reduction in the government's holding in public sector equity will not have an adverse effect on the health of the public sector units. On the contrary, the removal of government interference should give these units the corporate freedom they need to function efficiently in a competitive market.

Foreign trade policies in the Ninth Plan must be tailored to the objective of accelerating growth in an environment in which the world is becoming increasingly integrated and globalised. The process of globalisation is a reality which cannot be denied and also should not be avoided. However, it needs to be managed so that we can derive the maximum advantage from world markets. To do this, it is necessary to continue the process of opening up of the economy to international competition, which was initiated in the Eighth Plan period, while making parallel efforts to strengthen the potential of Indian industry to compete effectively in world markets. Quantitative Restrictions(QR) on imports have already been dismantled for all except consumer goods, and these remaining restrictions will be removed in a phased manner by the year 2003. As a special measure, QRs have been abolished in advance for SAARC countries, in order to given an impetus to regional trade. Import tariffs have also been reduced significantly over time, but our import tariff rates continue to be much higher than in other developing countries. Continuing with high levels of protection is not desirable if we want our industry to be competitive in world markets and it is therefore necessary to continue the process of phased reduction in import tariffs to bring our tariff levels in line with levels prevailing in other developing countries. Indian industry has adjusted well to the gradual reduction in protection levels over the past several years and the associated shift to a more realistic market determined exchange rate. The process has certainly helped to modernise industry and reduce costs and should be continued. Priority in tariff structure reform should be given to reducing tariffs where they cause tariff anomalies i.e. where duties on inputs are higher than on finished goods. Steps should also be taken to strengthen the anti-dumping machinery to ensure that the domestic industry is not subjected to unfair competition.

Export performance will be critical in determining our ability to achieve high growth rates while also achieving the objective of self-reliance. Rapid economic growth requires larger volumes of investment and extensive industrial modernisation, both of which imply rapidly growing import needs. Maintenance of balance of payments stability in this situation depends upon our ability to achieve high rates of export growth. Without a strong export performance there is a danger of being pushed into unsustainable balance of payments deficits, which could lead to increased external indebtedness. Export growth was relatively strong through most of the Eighth Plan period but has slowed down in the early years of the Ninth Plan. This is a disturbing development which needs urgent attention. It is essential to examine the constraining factors which limit our export performance and take urgent steps to remove these constraints. Exchange rate policy is one important determinant of export performance and it is essential to ensure that the exchange rate remains supportive of the export effort. Equally important in the present situation are infrastructure constraints including transport, customs procedures, and banking facilities. Our facilities and procedures in these areas are inadequate and often lead to delays and harassment. They need to be thoroughly revamped to provide greater support to exporters.

Foreign investment provides external resources which help to finance the balance of payments deficit without adding to the country's external debt. In this respect, the ability to attract foreign investment promote the objective of self reliance by helping to avoid a build up of external debt which adds to our vulnerability. Apart from providing support for the Balance of Payments, foreign investment also provides critical access to technology and other types of know how, and also provides potential linkages to world markets. In a world where trade is increasingly dominated by transnational corporations, it is important to encourage foreign investment as part of the process of modernising our industry and developing linkages with the rest of the world. Many Indian companies are eagerly seeking foreign investment in joint ventures as part of their plan for technological upgradation and modernisation and these efforts should be encouraged. As the second largest developing economy in Asia, India can be a major destination for foreign investment and policies in the Ninth Plan should be designed to take advantage of this possibility. Foreign investment policy was re-oriented in the Eighth Plan period to encourage a larger flow of foreign investment into the economy, especially into the infrastructure sectors where a rapid expansion in capacity is urgently needed. Foreign direct investment increased from $ 0.1 billion in 1990-91 to $ 3.2 billion in 1997-98. This level can be raised threefold by the end of the Ninth Plan period.

An aspect of external economic policy which needs careful consideration is the pace of liberalisation of the capital account. Globalisation has meant an explosion of financial integration in world markets, with large volumes of capital moving freely across national boundaries. Free access to capital from world market has its advantages, but it also poses dangers to developing countries because of the potential volatility of these flows. This is amply demonstrated by the experience of East Asian countries, which suffered a massive outflow of capital based on a sudden change in international perception about underlying fundamentals in these countries. The outflow precipitated a currency crisis with a severely disruptive effect on these economies. It is important to recognise that East Asia's problems arose not because of open trade policies or openness to foreign direct investment, but because an open capital account allowed banks and corporations to borrow freely in world markets leading to a large accumulation of short term debt. Weaknesses in the banking sector led to poor inter-mediation of these flows to support unproductive investments, including investment in real estate. When these weaknesses became apparent, the short-term nature of the liabilities led to a reversal of capital flow which was not easily arrested. A lesson commonly drawn from the experience is that developing countries with weak banking system may be well advised to maintain restriction on short term capital flows to limit the potential danger of sudden and large outflows. The approach to capital account liberalisation followed in India has been cautious, and government policy has been particularly concerned to avoid a build up of short term external liabilities. This caution is vindicated by the East Asian experiences. External policies should continue the process of cautious liberalisation in the financial sector with particular caution on the build up of short-term debt.

Infrastructure especially electric power, telecommunications, railways, roads and ports has become a major constraint on growth partly because there was inadequate investment in this area in the Eighth Plan. Rapid growth in the Ninth Plan will generate a heavy demand for these services, and being non-tradable the additional demand has to be met by expanding domestic supply. This in turn requires a large increase in investment. Traditionally, infrastructure has been provided largely by the public sector but the investment needs of infrastructure development greatly exceed the resources likely to be available with the public sector. Mobilisation of private investment to supplement the public sector effort is therefore essential.

However, it must not be assumed that private investment in infrastructure provides an easy solution to all problems justifying neglect of public investment in this area. Even with the maximum feasible success in attracting private investments, it will be necessary for the public sector to bear the brunt of the burden of expanding capacity in the Ninth Plan period. This is especially so in areas such as roads, railways and rural electrification, where the scope for private investment is limited, but it is also true for infrastructure in general that the public sector will remain the dominant source of expansion.

It is therefore critical to ensure that the resources available with the public sector are adequate to the task. This calls for far reaching reform of the infrastructure sector. Reforms of the power sector in the States, including reform of power tariffs, is especially important to make the State Electricity Boards financially viable. This is necessary to expand public investment and also to create credibility among private investors selling power to State Electricity Boards (SEB). The financial position of the Railways also needs to be strengthened. In both cases, apart from achieving higher operational efficiencies, a large part of the answer lies in introducing higher user charges. In other areas such as roads it is more difficult to rely on user charges though a start can be made by introducing tolls on public highways being expanded or improved. Additional financing can however be provided by an earmarked cess on diesel.

Public investment in infrastructure will have to be supplemented by efforts to attract private investment in these areas wherever possible. All major infrastructure sectors have already been opened to private investment and a number of private sector projects are being implemented in power generation, telecommunication services, ports and even roads and airports. However, these efforts have been less successful than was initially expected. Several projects have run into unanticipated difficulties which have delayed financial closure or slowed down implementation. In retrospect, it can be said that the difficulties and complexities of inducting private investment into regulated infrastructure sectors were underestimated. In the case of the power sector, the financial weakness of the SEBs has made it difficult for the independent power producers to obtain financing as long as they are subject to the risk of non-payment for electricity sold to the SEBs. The difficulty in negotiating fuel supply contracts with suitable penalties for non-delivery is also a problem. In the case of telecommunications the problems that have arisen are very different. Private investors bid high licence fees but subsequently claimed that the projects had become unviable with such high fees. Questions have also arisen about whether the policy framework provides a level playing field between new entrants and the existing public sector supplies. There can be no general solutions to these problems. However it is clear that if we want private investment in infrastructure we must ensure that the policy framework in each sector makes it possible for private investors to invest in these sectors with reasonable expectation of an attractive return. The deficiencies in the existing policies in each sector, as revealed by the experience of the past few years, need to be reviewed and urgently corrected. It should be the objective to put in place a policy framework which draws from best practice internationally and is tailored to the specific requirements of our situation.

Reforms in the real economy need to be supported by an efficient financial sector capable of moblilising the savings of the community and allocating them to the most productive users. In this context it is important to emphasise the role of financial sector reforms in the years ahead.

The Eighth Plan period saw the start of reforms in the banking sector consisting of financial liberalisation in the form of reduced direct Government control over interest rates and credit allocation mechanisms combined with efforts to strengthen the regulatory framework by improving prudential norms, capital adequacy standards and external supervision. While substantial progress has been made, the process of banking sector reforms needs to be accelerated, especially in the context of the recent experience of many East Asian countries where fragility in the banking sector was an important cause of the crises in the region.

Efforts must be made to ensure that the Indian banking sector comes up to internationally acceptable standards of capital adequacy, prudential norms and risk management. A process of reform has also been underway in the capital markets, and considerable progress has been made in establishing a modern regulatory framework covering the major participants in the capital market. The technology of trading and settlements in the stock exchanges has been modernised. This process also needs to be continued in the Ninth Plan. A strong well functioning capital market is an important part of the financial system and is a noteworthy important source of funds for both the private and public sectors in the future.

Another major area of financial sector reform relates to insurance and pension funds. These institutions for contractual long term savings can play a major role in increasing aggregate savings of the economy. Insurance and pensions institutions are also natural sources for long term capital since they have long term liabilities which they need to match with long term assets. The insurance and pension sector is therefore particularly important for the financing of infrastructure, which needs long term funds. Proposals for the reform of insurance sector by ending the public sector monopoly in this area, and opening it up to private sector participants, need to be implemented urgently in order to create a financial sector which provides good quality service to the consumer and also meets the financing needs of industry.

The success of any development strategy depends critically upon the ability of the Government to maintain a stable macro-economic environment in which the fiscal deficit is kept at a reasonable level. Unfortunately, this is an area where the experience of the Eighth Plan has been disappointing. The combined fiscal deficit of the Central and State Governments has remained between 6 and 7 per cent over the past several years, which is very high by international standards. Continued deficits at these levels will either mean continued high rate of interest, or a constant threat of over monetisation and consequent inflation. Steps must be taken to restore the fiscal health of the government at all levels. The Government borrowing programme must be contained at a moderate level if we want to move to a regime of low real interest rates with an adequate availability of resources to finance private investments at reasonable interest rates.

It is necessary to define a long term fiscal policy, with a clear time path for reducing the fiscal deficit to a sustainable level. Within framework, particular emphasis must be placed on reducing the revenue deficit so that containment of the fiscal deficit is not at the expense of public investment. Control over the revenue deficit can only be achieved if there is a willingness to face difficult decisions. Efforts must be made to improve tax collections especially in the area of excise duties where the tax ratio has been declining steadily. It is also necessary to take steps to control non-Plan expenditure. This in turn requires a systematic review of subsidies. Over the years, a number of subsidies, both direct and hidden, have come into existence. Fixation of administered prices on extra-economic considerations is an important element of non-transparent subsidies. Such subsidies are open-ended and in some cases have accrued to those who are not envisaged to be the real beneficiaries. There is a need to move to a system of transparent subsidies directed towards specific target groups in a selective manner.

The Eighth Plan has witnessed significant improvement in technology levels in the country, particularly in the industrial sector. Much of this development has been based on import of technology which is certainly important and needs to be continued. However, it is also necessary to emphasise development of technological capabilities within the country. Steps need to be taken to enhance the technological capabilities of the nation and to this end, it is necessary to evolve a technology policy which gives adequate incentive to the development of domestic R&D. A substantial domestic R&D capacity must be an essential component of our long run development strategy. This aspect of development strategy is further emphasised by the emerging global regimes restricting the flow of a wide range of technologies on the ground of "dual use." This situation has to be met through a combination of international initiatives and renewed domestic effort. Government must evolve strategies which will help domestic R&D activities through an appropriate combination of tax and other incentives. It is also necessary to strengthen the legal framework protecting intellectual property to give appropriate incentives for research and development.

The evolution in information technology taking place around the world holds for the prospect of considerable economic gains for India both in domestic application of IT and in our economic interaction with the rest of the world. The spread of information technology in India is as yet inadequate, and urgent steps are needed to create a national network for information dissemination. This would not only help our industrial sectors, and especially the software industry but would also enhance the effectiveness of dissemination of technologies for strengthening rural economic activities, including agriculture, and social infrastructure and would also enable greater transparency in the operation of Government schemes and programmes. India has the potential to become a software super power in the next ten years and it must be the objective of policy to realise this ambition. The government has already taken a number of steps in this area and it must be ready to do more.

The Ninth Plan is based on the concept of cooperative federalism whereby much greater freedom would be given to states to determine not only their own priorities but also the modalities of public intervention and provision of goods and services. This concept also embodies a higher degree of dialogue and coordination between the states. The central government's principal role in this regard would be to raise the issues which appear to be of national interest and to provide a forum for arriving at a consensus. These issues are dealt with in greater depth in Chapter6.

A disturbing development in recent years is the inability of the States to mobilise resources as targeted leading to a shortfall in the share of States in total Plan outlay. The share of the States in the Eighth Plan was only 36.4 per cent as compared to the projected share of 41.5 per cent. When the states' share declines, the sectors which suffer more severely are agriculture, basic minimum services, health, education, women and child development, welfare and also economic infrastructure such as electricity. In view of the emphasis that is being placed on the importance of these sectors, this trend needs to be halted and reversed. This is a relevant consideration in determining the support from the Central Government for State plans. However, the major correction has to come by reversing the deteriorating trend in States' resources for the Plan. This in turn focus attention on the growth of non-plan expenditure in the States and the very large increase in losses in sectors such as electricity, irrigation and road transport which seriously erodes States' resources.

The role and function of the Panchayati Raj Institutions (PRIs) has important implications for development strategy. Past experience has shown that functions and services which are localised in nature are most efficiently implemented by local political entities and other local bodies. It is therefore necessary to devolve not only greater resources and responsibilities to the PRIs, but also to grant greater powers to them for raising their own resources. In this context, programmes in the social sector which are of national importance, and therefore have a high level of central government involvement, but which are essentially localised in their delivery, would need to be operated through a more direct relationship between the central government and the PRIs in order to increase their effectiveness. This mechanism is already operational in certain centrally sponsored schemes (CSS), and needs to be extended to others as well. The transfer of some CSS along with the funds, as already initiated, is a step in this direction.

The Eighth Plan had identified peoples' initiative and participation as a key element in the process of development, particularly in improving the effectiveness of development outlays which has been declining over the years. It had also recognised that the role of the Government should be to facilitate the process of peoples' involvement by creating right types of institutional infrastructure, particularly in rural areas. The progress on this front has not been entirely satisfactory principally due to the fact that the other tiers of the Government were not fully integrated into the development strategy. The process of social mobilisation and development of peoples' initiatives cannot be achieved without the active support and involvement of the political system at all levels. The Panchayati Raj bodies in rural areas and Nagar Palikas in urban areas will have to be directly involved in the development process. People's involvement via their elected representatives will be realised through genuine democratic decentralisation.

Other forms of peoples' participation also need to be strengthened. From the early days of planning, cooperatives have been perceived as the most important form of peoples' institution for promotion of equity, social justice and economic development. Every effort will be made to make the cooperatives work. They need to be liberated from tight bureaucratic control. Self-help Groups, Associations of Workers or Small Producers, etc. are other forms of institutions which will be encouraged. Government will seek active partnership with the voluntary sector in organising and promoting these institutions.

The Indian planning process has always laid emphasis on measures to ensure sustainability of the development process not only in economic terms, but also in terms of social and environmental factors. Much of the measures adopted in the Agenda 21 of the United Nations Conference on Environment and Development already find reflection in the Indian plans. The Ninth Plan will need to carry this tradition further and consolidate on the creation of conditions for sustainable development. Rapidly growing population, urbanisation, changing agricultural, industrial and water resource management, increasing use of pesticides and fossil fuels have all resulted in perceptible deterioration in the quality and sustainability of the environment. It is recognised that environment protection does not only involve a prevention of pollution and of natural resource degradation, but has to be integrated with the over-all development process and the well-being of people.

AREAS OF SPECIAL IMPORTANCE FOR STATE INTERVENTION

The Ninth Plan views the role of the state and private sector as complementary and both are essential. Private sector activity needs well functioning markets and hence the emphasis on economic liberalisation and decontrol. However, while market forces are extremely important, there are many circumstances in which markets may not exist or, even if they do, may not work efficiently and effectively. There are also conditions under which unbridled operation of market forces may give rise to outcomes which may be deleterious when seen in a broader national and social perspective. Government regulation and intervention in markets becomes justifiable in such situations though this should not become an open-ended rationale for excessive intervention. Government interventions have to be strategic and must emanate from a vision of the role and responsibility of state policy and public action where markets are likely to be imperfect. Three broad areas are especially relevant in this context. These are: (a) quality of life of the citizens; (b) generation of productive employment; and (c) regional balance.

Quality of Life

Eradication of poverty and provision of basic minimum services are integral elements of any strategy to improve the quality of life. No developmental process can be sustainable unless it leads to visible and widespread improvement in these areas. There is by now enough evidence to show that rapid growth has strong poverty reducing effects and, given a public policy stance which is sensitive to the needs of the poor, a focus on accelerated growth will also help in realising the objective of alleviating poverty. The shift in emphasis towards private initiative and reliance on the entrepreneurial spirit of the people essentially seeks to create the conditions for rapid and sustained growth. Nevertheless there are aspects of growth which can be labour-displacing and impoverishing. These arise essentially out of unequal initial endowments of physical resources, human capital and information, which prevent segments of the society, particularly women and other socially and economically disadvantaged groups like the Scheduled Castes, Scheduled Tribes, Minorities, Other Backward Classes, disabled etc. from taking advantage of the opportunities that become available or from protecting their interests in an adequate manner. Much of these disadvantages have their roots in historical circumstances and are perpetuated through social and political factors. Market forces left to themselves may not correct these disparities, and indeed may accentuate them since there would very likely be a tendency to exploit the weak and the disadvantaged by economic alignments of the socially and politically advantaged. Public action will be necessary to reduce and eventually eliminate these schisms through a focus on removal of the initial handicaps and stringent protection of the rights of the disadvantaged.

It is important to emphasise that the process of elimination of historical disadvantages such as gender, caste and other types of social and economic inequalities should not be based on increasing dependency of these segments of the people on the government. Sustainability of the growth process demands that they should be viewed as active participants, and indeed as partners, in the process of development and not as passive beneficiaries of public largesse. The need therefore is to create the conditions by which the disadvantaged are not only empowered to take advantage of the opportunities created by the growth process but also to contribute actively in the process of creation of wealth and well-being. For this purpose, the individual capabilities of the people have to enhanced through education, information and access to appropriate technology. Amelioration of the immediate deprivation through anti-poverty programmes should be viewed only as a transitional arrangement, and every effort directed towards ensuring that these can be phased out at the earliest.

Empowerment of the historically disadvantaged will require more than provision of the basic capabilities for integrating them into the growth process. Appropriate institutional structures will need to be created and encouraged in order to allow full play to their productive and entrepreneurial energies. The objective conditions of the Indian economy are such that the vast majority of the populace will continue to have to be engaged in self-employment or casual employment in the foreseeable future. Thus any effort at raising the growth rate of the economy through increases in productivity and entrepreneurial dynamism would need to look beyond the usual forms of production organisation. In order for such activities to be viable and remunerative in a market framework, alternative forms of organisation will have to be recognised and nurtured. Revitalisation of the cooperative sector and other forms of economic association of people is of great significance in this context. Cooperatives have played an important role in development and promotion of equity and social justice. Some have grown into substantial size, but there continue to be policy and procedural limits on their growth and diversification. There is a need to de-bureaucratise and de-politicise the operation of this sector and to enable it to access resources from other sources on its own strength.

In the past, food and nutritional security has been largely interpreted to mean adequate availability of basic food products in the country as a whole. The concept of food security now needs to be broadened to include peoples' access to basic nutritional requirements, both physically and economically. This problem is particularly acute in the vulnerable sections of society and in the deficit and inaccessible regions of the country. It is necessary to develop strategies by which such inadequacies can be overcome by integrating the food production and distribution systems with the employment and poverty alleviation programmes. In particular, the Public Distribution System (PDS) will need to be restructured in order to provide foodgrains at substantially lower prices to the poor in a focused manner and to ensure availability of such commodities in the remote and deficit areas of the country. A wide-spread, well-established PDS is essential for ensuring food security.

Economic growth and employment opportunities in themselves may not be sufficient to improve the living conditions of the poor. They need to be accompanied by measures which enhance the social and physical conditions of existence. Despite considerable efforts, provision of social infrastructure and services remain inadequate. These are areas which are in the nature of quasi-public goods and in which private initiatives and the market are unlikely to play a significant role. Primary education, primary health care, including the preventive and promotive, safe drinking water, nutrition and sanitation require heavy investment which has to be provided out of public funds. However, since the requirements of social infrastructure vary significantly across regions, greater decentralisation of decision-making than exists at present is desirable. Furthermore, recognising the localised nature of these essential services, it is desirable that the control over the operation and maintenance of the facilities should be in the hands of peoples' institutions and local associations, with adequate resources being made available either from the exchequer or through devolution of powers to raise such resources.

In recent years the problems of rapid urbanisation has become acute. There has been a progressive decline in the availability of essential services as well as in the quality of life in urban areas. The urban poor have been the worst affected segment in this process of decline. The health and environmental consequences of increasing population density, lack of safe drinking water and inadequate urban sanitation are likely to become further aggravated unless steps are initiated during the Ninth Plan to improve the situation through a well considered and articulated urbanisation policy with identified programme components including those for disease surveillance, epidemic control and urban solid and liquid waste management.

Social and demographic indicators cannot improve merely through increased investment, but in addition require a significant change in social attitudes and behavioural responses of the people. In order to achieve these objectives, there is no alternative to social mobilisation and community participation. In this process the role of women is critical. The process of empowerment of women at the political level has already begun, but it needs to be carried forward into the social and economic spheres as well. Special emphasis would have to be placed on ensuring that control of social infrastructure, particularly in health and education, in the public domain is vested in women and women's organisations.

A synergy between environment, health and development needs to be explicitly recognised. No development process which leads to an adequate quality of life can be sustained in a situation of deteriorating environmental and ecological conditions. Environmental degradation is usually the outcome of individual actions which do not take into account the externalities imposed on others both in space and time. The market mechanism left to itself does not provide any method of forcing the internalisation of these costs. Indeed, it may actually reinforce such behaviour. As a consequence, it is the responsibility of the government to ensure that the interests of both the current victims of environmental degradation and future generations are adequately protected, without disrupting the process of growth and development. Eventually, the responsibility for preservation of the environment rests on the people themselves. In order for these responsibilities to be properly discharged, however, the rights to a healthy environment have to be unambiguously assigned through legislative action; and people's participation in preservation of the environment fostered through social mobilisation and appropriate organisational structures.

Finally, all efforts to improve the quality of life can flounder under the pressure of population growth. This is another area where immediate individual interest may be in conflict with both the national interest and the longer-term interests of individuals. While there is evidence that rapid growth and development does generate positive effects towards population containment, the pace of such reduction may not be sufficient in view of the pressures that are being experienced on the nation's limited natural resource base. Concerted public action, both direct and towards awareness-building, is essential for creating the conditions by which the country can stabilise its population at a level that can be sustained in the long-term.

Employment

Productive work is not merely a means to the ultimate end of economic well-being. It should be seen as an end in itself. It is a basic source of human dignity and self-respect. It is also an integral element in nurturing national identity and social cohesion. Although rapid economic growth is the surest and most sustainable source of increasing work opportunities, market forces alone cannot be relied upon to provide gainful work for all. This is particularly true in a labour-surplus economy like India. The congruence between the skill and spatial distribution of the labour force and the corresponding requirements of the economy may not be achieved through market-based mechanisms due to numerous distortions in the labour markets, information systems and infrastructural endowments that continue to persist in the country.

A primary objective of state policy should be to generate greater productive work opportunities in the growth process itself by concentrating on sectors, sub-sectors and technologies which are more labour intensive, in regions characterised by higher rates of unemployment and underemployment. The experience of the last decade has been that relatively rapid growth does generate adequate employment opportunities. However, there is no cause for complacency. The magnitude of labour force growth in the coming decade is the highest that it has ever been or is ever likely to be. The skill and geographical distribution of the emerging labour force reflect past trends in population growth and skill development, and cannot be changed significantly at this stage. Public intervention will be necessary to ensure not only that adequate work opportunities are created, but also that the labour force is able to access these opportunities.

The agricultural sector has historically played an extremely important role in providing employment in India, and is likely to continue to do so at least for the next decade or more. The skill profile of the emerging work-force is such that a large majority will have to continue to seek work opportunities in this sector. In order to create such opportunities, land and tenancy reforms are essential. The evidence strongly suggests that small farmer agriculture, based on secure rights on the land, is not only more productive per unit of land than either large farms or where the rights of the actual tiller are not well defined, but is also more labour absorbing. Active government intervention, both in implementation of existing laws and formulating new legislation for ensuring the tillers' rights, in these areas will have to be taken up in earnest in order to create conditions of high labour absorption with adequate productivity growth.

In a dynamic economic environment, restructuring or failure of firms and even demise of certain economic activities is natural and to be expected. Efforts at preventing such adjustments and failures in the interest of protecting employment has proven to be both inefficient and ineffective. The welfare effects of such measures in terms of the trade-off between the interests of the currently employed and those of either the unemployed or the new entrants to the labour force are also not unambiguous. While providing avenues for retraining, skill upgradation and reemployment, the focus of public policy needs to be shifted from protecting specific jobs to protecting the interests of the work-force. This has two dimensions. First, provision has to be made for the payment of the legitimate dues of and for mitigating the hardships faced by the displaced work-force. Such systems by and large exist for the organised sector, but the labour force in the unorganised sectors and the self-employed are presently uncovered. Providing such support is not easy to do without the creation of appropriate people's organisations with active government support. Second, all efforts have to be made to find alternative occupation for the unemployed. Although the minimum necessary condition for ensuring such reemployment is that the growth rate of creation of work opportunities must exceed the growth rate of the labour force, it is not sufficient. Existing workers embody a certain structure of skills which may not be appropriate for the new work opportunities that may be created. The market by itself may not be prepared to impart these skills to the displaced workers, since the preference will generally be for training the younger, new entrants in the labour force, whose potential life-time returns to the investment in human capital is likely to be significantly higher. It is necessary, therefore, to establish systems in the public domain for retraining or reskilling the displaced labour force in line with the requirements of the economy.

Even if the generation of work opportunities is adequate, the quality of employment in terms of the incomes received and the work environment may leave much to be desired, and may be likely to remain so without directed public action. It is therefore necessary to shift the focus of employment strategies towards creating conditions whereby employment opportunities lead to significantly better living and working conditions of the people and to uphold the dignity of labour. In particular, the incidence of scavenging and child labour which arise out of acute poverty and prevailing social disabilities need to be eradicated keeping in mind the requirement of maintaining family incomes. Public investment in modern sanitation systems, on one hand, and focused provision of appropriate education and skill, on the other, are necessary to tackle the first social evil. In so far as the second is concerned, precipitate action may not always achieve the objectives. A family-centric strategy and sensitivity to the needs of the orphan or run-away, coupled with rigorous implementation of the existing laws, will need to be evolved to effectively tackle this problem.

It is not enough to merely create the right kinds of employment opportunities, but also to provide the people with the human capital by which they can take advantage of these opportunities. Education and skill development are the essential features of such empowerment. Free and compulsory education for children, especially for the girl-child and other educationally backward groups, supported by an adequate mid-day meal programme in schools is the first step towards this end. Vocationalisation of the education system is essential, particularly since there is evidence that there are a number of skills which are in acute short supply and for which there is a serious shortage of training facilities. In addition, special programmes will have to be implemented to develop skills, enhance technological levels and provide marketing channels for people engaged in traditional occupations. The education and training systems in the country if not reformed will begin to display a certain degree of rigidity and get increasingly divorced from the needs of the economy and the social infrastructure. This trend needs to be reversed, and efforts will have to be made to introduce a system by which the emerging skill requirements of the economy are anticipated and the education and training systems reoriented accordingly. Although the private sector, which will be the main source of work opportunities in the future, will have an important role to play in this regard, the initiative will have to be taken by the government since the bulk of the educational system in India is in the public domain.

Recognising the high incidence of underemployment and increasing casualisation of labour, there is need to enhance employment opportunities for the poor, particularly for those who are in seasonal occupations. The recent trend in lower labour force participation rates among women is partly a reflection of the inadequacy of appropriate work opportunities. In view of the observed relationship between women's participation in the labour force and fertility, this trend has disquieting portents regarding the pace of reduction in the growth rate of population. There is need therefore for public intervention in creating work opportunities which are sensitive to the seasonal and locational needs of the underemployed, particularly women. Such opportunities should be demand-driven to the extent possible. In this context, the effort to implement a national Employment Assurance Scheme is of considerable significance.

Regional Balance

Balanced regional development has always been an essential component of the Indian development strategy in order to ensure the unity and integrity of the nation. Since not all parts of the country are equally well endowed to take advantage of growth opportunities, and since historical inequalities have not been eliminated, planned intervention is required to ensure that large regional imbalances do not occur. With greater freedom and choice of location that is now available to industry, it is more than likely that some states would be able to attract more private investment than others. In such a situation it will be necessary to deliberately bias public investment in infrastructure in favour of the less well-off states. It will also have to be ensured that the states which benefit from this reorientation do not dilute their own efforts at generating investible resources or divert their resources to other uses.

Efforts at attracting private investments by state governments has already led to a certain degree of competition in granting of fiscal and other concessions. This tendency is likely to be accentuated in the future as private investment becomes increasingly more important. While a certain degree of competition in this regard may be desirable, in the long-run it would adversely affect the fiscal viability of some of the states and thereby jeopardise the ability of these states to provide the basic social and economic infrastructure. This would have serious implications regarding the future progress on regional balance and would tend to accentuate regional disparities in quality of life. It is necessary therefore for states to operate in a spirit of cooperative federalism and to arrive at a set of public policy and action in which state-level initiatives at attracting private investment in a competitive manner will be acceptable, and those in which a common position would be taken by all states in their collective interest.

There are also regions or pockets of poverty which are unable to benefit adequately from the over-all growth process. For the most part such pockets of poverty reflect the inadequate integration of the local economies in the wider growth process due to historical reasons. There is a strong probability that even in the future, market-based growth may elude these regions unless active intervention is made by the government. Sustainable anti-poverty programmes in such regions would need to involve not only direct employment and income enhancing policies to tide over the immediate deprivation, but also measures by which the asset endowment of the poor can increase and such areas can become better integrated with the rest of the economy for sustainable growth.

The focus regarding the issue of backwardness and regional balance has traditionally been on industrialisation. The evidence, however, suggests that reduction in regional disparities, particularly in average standards of living, may be better achieved through greater focus on agriculture and other rural activities. For this it is necessary to not only increase the productivity of agriculture in backward areas, but also to increase the degree of integration between the rural areas and the rest of the country through improved connectivity in terms of transport and communications, and provision of marketing support. In the absence of adequate public investment in such infrastructure, the prospects of a disparity-reducing agricultural growth will recede.

In some regions of the country, relatively high levels of income continue to be associated with low human development indicators and poor provision of social infrastructure, such as safe drinking water, primary health and primary education facilities. These are areas in which private initiatives are unlikely to play any major role, and it would be the primary responsibility of the government to ensure that appropriate interventions are made. There is therefore a need to identify the minimum norms of availability of such infrastructure and to accelerate public investment for achieving these norms in a balanced and timebound manner. The problem of shelter for the poor, both in urban and rural areas, is particularly acute and a programme of providing assistance for construction of houses will need to be implemented.

The issue of regional balance operates at both the inter-State and the intra-State level and there is therefore the need to address it in a framework that is more flexible than the political and administrative boundaries of each individual state. Growth and development are intimately related to the level of economic integration of the various parts of the country and the linkages that are formed between the backward and the more developed regions. In view of this, it is necessary to move away from the concept of competitive policy formulation to a framework of cooperative federalism, wherein neighbouring states adopt a common set of strategies for development of their backward areas in a coordinated manner and remove or reduce the barriers to inter-state commerce that have been built up over the years. Cooperative action can not only prevent loss of revenues to state and local governments that may arise from the reduction of existing fiscal barriers, but actually augment the flow of resources.

In view of the resource constraints being faced by the Government at all levels, the prioritisation of the various facets of quality of life would have to be carried out on a region-specific basis. In particular, it will be necessary to identify those areas of the country where the growth process will more or less take care of the problem of acute deprivation. The focus of anti-poverty programmes would have to be shifted to those other areas which are as yet inadequately benefited by the growth process. In such areas the pre-conditions for a market-based development strategy will have to be created expeditiously, even if it means a certain degree of positive bias.

THE ROLE AND METHODOLOGY OF PLANNING

The Planning Commission believes that the principal task of planning in a federal system is to evolve a shared vision of and a shared commitment to the national objectives and the development strategy not only in the Government at all levels but also among all other economic agents. No development strategy can be successful unless each component of the economy works towards a common purpose with the full realisation of the role that has to be played within an overall structure of responsibilities. The principal function of the approach to the Ninth Plan is not only to articulate such a shared vision, but also to ensure shared commitments.

Based on such a shared vision, the principal task during the Ninth Plan is to build on the successes of the Eight Plans, while tackling the problems that have emerged, particularly in areas such as capital formation in agriculture, living standards of the poor, infrastructure, social sector, regional disparity and fiscal deficits. Despite the considerable progress that has been made by the Indian economy in both economic and social spheres, the development task is far from complete and resources continue to be limited. Moreover, as has been indicated, the Indian economy is still vulnerable and the operation of a more open economic system has to be tempered by judicious public interventions to ensure that these vulnerabilities are gradually overcome.

Recognising the federal nature of the Indian system, the planning process has to develop a common policy stance which would be adopted both by the Centre and the States. The role of planning would therefore involve considerable degree of policy coordination between the Centre and the States, between the States, and between the States and the sub-State level tiers of Government.

With the progressive deregulation of the economy, and the larger role of the private sector and market based decision making, planning methods based on input output balances for each industrial sector have become less relevant. However, sector level planning for the infrastructure sectors and broad planning for critical sectors such as agriculture remain highly relevant both to identify inter-sectoral constraints and as a guide to government policy action. The planning process also needs to take stock of the resources for development to the internal and external, and to indicate methods by which these resources can be augmented in a sustained manner. Since planning is done for the entire nation, the strategies for resource augmentation need to cover not only the government at all levels but also those which would be available to other sections of people. The other principal task of planning, particularly in a market oriented economy, is to identify the areas of emerging vulnerabilities and to suggest measures to address them. Although the specific policies would need to be worked out by the concerned ministries or the States, the broad directions would have to be provided by the planning system. Unless such problem areas are placed within the wider macro-economic context, short-run fire-fighting measures can lead the economy in undesirable directions.

With the reduction in the instruments available to Government in controlling economic matters and the greater degree of uncertainities that are present in any market based system, the methodology of planning will have to change if it is to retain its relevance. The Eighth Five Year Plan had explicitly indicated a shift from directive to largely indicative plan. There is considerable variance in the interpretation of this term. The Ninth Plan is based on a more specific modality of planning which involves working out of a consistent and desirable development path, the identification of emerging trends and deriving policy measures to bring about a confluence between the two. The planning process today, therefore, focuses on planning for policy so that the signals that are sent to the economic system induce the various economic agents to behave in a manner which is consistent with the national objectives.

The most important characteristic of the methodology adopted in the formulation of the Ninth Plan is that it is not based on a deterministic relationship between the Plan and economic performance. It is explicitly recognised that there are uncertainities in the system and limitations in the ability of the planning system to accurately predict future trends. Furthermore, it is also recognised that the effects of Government policies and interventions are not entirely predictable. This is certainly true of economic variables, although it may not be so for the performance in social sectors such as health and education. The planning approach adopted for the Ninth Plan therefore concentrates on pointing out the likely outcomes and suggests the major directions for policy intervention. The details of the actual policies and the manner of implementation will need to be worked out from time to time between the concerned Ministries, States and the Planning Commission.

DEVELOPMENT PERSPECTIVE AND SUSTAINABILITY OF GROWTH

Development and growth are long term processes, and a five year plan represents only the characterisation of the economy and social processes over a period in which a fair degree of reliance can be placed on estimates and projections. The viability and sustainability of the development strategy embodied in a five year plan needs to be seen in the longer perspective in order to ensure that continuity of directions and the balance between interests of the present and the future generations is maintained. The Indian Five Year Plans have traditionally been embedded in a 15 year perspective period. The Ninth Plan continues this tradition.

Since the principal objective of economic planning is to fulfill the social and human aspirations of the people, the basis for all planning efforts rests upon the projections of future demographic trends, the requirement for work opportunities and the need to erradicate poverty. The factors which determine the long term conditions of growth, along with the size and skill profile of the work force, are the availability of investible resources through domestic savings and prudent access to international investment funds, the basic resource endowment of the country, the available and latent entrepreneurial abilities of the people and the pace of technological development.

POPULATION GROWTH & DEMOGRAPHIC PROFILE

Population stabilisation is an essential prerequisite for sustainability of the development process. The National Family Planning Programme was launched in 1952 with the objective of "reducing birth rate to the extent necessary to stabilise the population at a level consistent with requirement of national economy". The technological advances and improved quality and coverage of health care resulted in a rapid fall of mortality rate from 27 in 1951 to 9.8 in 1991. In contrast, the reduction in birth rate has been less steep, declining from 40 in 1951 to 29.5 in 1991. As a result the annual exponential population growth has been over 2% in the last three decades. The report and recommendations of the NDC Committee on Population gave a new thrust and dynamism to the Family Welfare Programme during the Eighth Plan period. As a result of these recommendations, the centrally defined method-specific targets for Family Planning have been abolished and decentralised area-specific need assessment to meet all the felt needs for contraception, maternal and child health have been initiated.

Indeed this was the basis on which one half of the funds under the social safety net scheme was provided to districts with crude birth rate (CBR) over 39/1000 (as per the 1981 Census) to improve existing facilities for safe delivery and child care in these districts. During the Eighth Plan period the fall in birth rate has been steeper than that in the death rate; consequently annual growth rate is around 1.9% during 1991-95. The projections of the major demographic indicators for the perspective period are presented in Table 1-1. It may be seen that the rate of decline in population growth is likely to be accelerated during the Ninth Plan period and will continue to do so during the perspective.

Although the above projections made by the Technical Group on Population Projections constituted by the Planning Commission in 1996 are based on well-accepted and time-tested methodology and should have a high degree of accuracy, they are nevertheless projections. Any such exercise is based on extrapolation of past trends and behaviour, and can never adequately capture possible departures from the trend. Reproductive behaviour is affected by an extremely complex set of economic and social variables. Although it is difficult to derive precise relationships between these variables and the rate of population growth, some of the linkages are well accepted in qualitative terms. In particular, factors such as female literacy, infant mortality rates, average age at marriage and labour force participation rate for women are believed to exert considerable influence on fertility. It is quite likely therefore that the special emphasis that is being placed on basic minimum services and on empowerment of women in the Ninth Plan may lead to sharper increase in these variables and thereby to even more impressive fall in the growth rate of population than indicated in the table.

Table 1-1 : Major Demographic Indicators

	1996	2001	2006	2011
1. Population (million)	934.22	1012.39	1094.13	1178.89
2. Urban Population (%)	27.23	28.77	30.35	31.99
3. Sex Ratio	107.9	107.2	106.6	106.0
(males per 100 females)				
		1996-2001	2001-06	2006-11
4. Growth rate of population(%p.a.)		1.62	1.57	1.50
5. Expectation of life				
at Birth (years)				
Male		62.36	63.87	65.65
Female		65.27	66.91	67.67

Note :

1. Population as on March 1.

2. Expectaion of life at birth is estimated from SRS based Abridged Life Tables, 1989-93, Registrar General, India.

SOURCE: Report of the Technical Group on Population Projections constituted by Planning Commission, published as: Population Projections for India & States 1996-2016, Registrar General, India (1996)

One of the features of Indian demographic behaviour which has been the cause for considerable concern has been the more or less steady decline in the sex ratio during the current century from 972 females per thousand males in 1901 to 927 in 1991. The factors responsible for this continued decline are as yet not clearly identified. However, it is well recognised that the adverse sex ratio is a reflection of gender disparity and appropriate steps to correct this trend will need to be taken during the Ninth Plan period. It is expected that the steps that are being proposed will lead to an improvement in the sex ratio to about 943 by the end of the perspective period. A major component in the reversal of the sex ratio trend would be a substantial reduction in maternal mortality. At present the longevity at birth of women is only marginally higher than that of men. However, over the next decade the difference in life expectancy between men and women is expected to progressively increase.

Table 1-2 : Rate of Growth and Share of Projected Population

States	Rate of Growth (1997-2012)	Share of population to total 1997	2012	Year by which TFR=2.1 may be achieved
Andhra Pradesh	1.06	7.48	6.98	2002
Assam	1.39	2.65	2.58	2015
Bihar	1.78	9.71	10.07	2039
Gujarat	1.37	4.73	4.62	2014
Haryana	1.54	1.94	1.94	2025
Karnataka	1.24	5.13	4.91	2009
Kerala	0.90	3.21	2.92	1988
Madhya Pradesh	1.77	7.74	8.01	2060
Maharashtra	1.09	9.01	8.44	2008
Orissa	0.89	3.58	3.25	2010
Punjab	1.15	2.33	2.20	2019
Rajastan	1.82	5.20	5.42	2048
Tamil Nadu	0.82	6.17	5.55	1993
Uttar Pradesh	2.22	16.37	18.10	2100
West Bengal	1.29	7.77	7.50	2009
Rest of India	2.05	7.03	7.58	
All India	1.54	100.00	100.00	2026

SOURCE: Report of the Technical Group on Population Projections constituted by Planning Commission, published as: Population Projections for India & States 1996-2016, Registrar General, India (1996)

The average growth rate of population during the 15 year perspective period of 1.54 per cent is distributed extremely unevenly between the different states of the nation. The statewise estimates of the likely rates of growth of population are given in Table 1-2. As may be seen, there is a wide variation in the growth rates, varying between 0.89 per cent per annum in Orissa to 2.22 per cent per annum in Uttar Pradesh. As a consequence of these differential population growth rates, the share of different states in the total population of the country is likely to change quite substantially over the perspective period. In particular, the population share of northern states of Bihar, Madhya Pradesh and

Uttar Pradesh are likely to increase substantially, whereas almost all other States would either have a constant or declining population proportion. A matter of greater concern relates to the goal of population stabilisation. In recent years, the earlier approach of using the Net Reproduction Rate of unity (NRR = 1) as the overall indicator of the replacement level of fertility has been changed to a Total Fertility Rate (TFR) of 2.1. The projected level of the TFR for the country as a whole in 2012 is 2.52, which is significantly higher than the level required for attaining the replacement rate, and the desired rate may not be achieved until 2026. As can be seen, if the country has to attain the desired replacement rate within the perspective period, urgent measures at fertility reduction would need to be taken in Bihar, Madhya Pradesh, Rajasthan and Uttar Pradesh.

Though the decline in birth and death rates have occurred in all states, the rate of decline was slower in some states like U.P. and Bihar; even within the same state there are substantial differences between districts. The Family Welfare Programme, therefore needs to be re-oriented to remove or minimise the inter- and intra-State differences, with emphasis on improved access and quality of reproductive and child health services through PHC-based decentralised area-specific microplanning without recourse to specific centrally fixed targets. It is imperative that adequate data-base is available at the PHC level on a yearly basis, both for need assessment and also to provide an independent mechanism for impact assessment and midcourse correction. This can be achieved only through complete registration of all births and deaths. For this, it is necessary to strengthen the Civil Registration system through involvement of ICDS workers, Panchayati Raj institutions, as well as health personnel.

Simultaneously, the Ninth Plan will launch an intensive drive to promote health education, particularly in reproductive health, so that India builds a sound foundation for a successful preventive and promotive national health paradigm. Basic principles of hygiene, sanitation, nutrition, and prevention of illness and disease will be promoted through not only the educational institutions and the adult education programme but also through the ICDS programme, through counselling offered by the health workers at all levels, and the mass media.

Table 1-3:Important Social & Demographic Indicators-Major States

States	Birth Rate (1996)	Death Rate (1996)	Infant Mortality Rate (IMR) (1996)	Couple Protection Rate (CPR) (1995)	Total Fertilty Rate (TFR) (1994)	Life expectancy females (89-93)	Female Literacy Rate (1991)	Mean age at Marriage (females) (1994)
(1)	(2)	(3)	(4)	(5)	(6)	(7)	(8)	(9)
Andhra Pr.	22.7	8.3	66	46.9	2.7	61.5	33.7	17.8
Bihar	32.1	10.2	72	21.1	4.6	57.2	23.1	18.6
Gujarat	25.5	7.6	62	57.4	3.1	61.1	48.5	20.4
Haryana	28.8	8.1	68	53.9	3.7	63.7	40.9	19.2
Karnataka	23.0	7.6	53	55.6	2.8	63.5	44.3	19.4
Kerala	17.8	6.2	13	46.7	1.7	74.7	86.9	22.3
Madhya Pradesh	32.4	11.1	97	47.4	4.2	53.8	28.4	18.8
Maharashtra	23.2	7.4	48	51.0	2.9	65.4	50.5	19.1
Orissa	26.8	10.7	95	39.5	3.3	55.3	34.4	19.5
Punjab	23.5	7.5	52	76.9	2.9	67.6	49.7	20.3
Rajastan	32.3	9.1	86	32.6	4.5	58.5	20.8	18.4
Tamil Nadu	19.2	7.9	54	51.7	2.1	63.4	52.3	20.2
Uttar Pradesh	34.0	10.2	85	37.2	5.1	55.1	26.0	19.5
West Bengal	22.8	7.8	55	34.2	5.0	62.3	47.1	19.5
All India	27.4	8.9	72	45.4	3.5	59.7	39.4	19.4

Source: Col. 1,2,3,5,7 & 8 SRS Estimates Col 4 Deptt of Family Welfare Col 6 SRS Based Abridged Life tables, 1989-93, RGI, India

While demographic transition reflects quantitative and qualitative changes in the population profile, the on-going changes in disease burden is producing a major health transition as reflected in the wide difference between the present and projected distribution of disease burden expressed as a percentage of disability adjusted life years (DALYs) by broad cause group. Communicable diseases, along with maternal and perinatal morbidity, presently constitute 56.4 per cent of DALYs, while non-communicable diseases constitute less than 30 per cent. Over the next 15 years, non-communicable diseases are likely to constitute nearly half of the distribution of total disease burden, thus posing a serious future threat to the health of the population. While a decline in communicable diseases as a result of public health measures would be a welcome development, disproportionate increase in non-communicable diseases such as high blood pressure, diseases affecting the heart, diabetes and cancer would be in large measure due to the rapidly changing life styles, both in terms of quantity and quality of food consumed and increasing level of physical inactivity. Preventive and promotive measures need to be urgently initiated to avoid the projected outcomes, in addition to incorporating within the health system, community education and early detection as well as management of such diseases.

Despite the steady reduction that is expected to occur in the rate of growth of population, the biggest challenge facing the Ninth Plan, and indeed the entire perspective period, arises out of the projected age structure of the population that is emerging from the past demographic trends. These projections are presented in Table 1-4. It will be seen that the growth rate of the population between the ages of 15-60 is likely to be considerably higher than the growth rate of population. Since this is the age group which is likely to be actively seeking employment, the need for expanding work opportunities will be considerably higher than before. On the positive side, the total dependency ratio is likely to decline quite significantly over this period. This occurs through a sharp reduction in the dependency ratio pertaining to the young, with that for the old remaining more or less constant. Therefore, if gainful employment can be provided to the entire work-force, the economic conditions of the households should improve significantly faster than earlier.

Table 1-4 : Projected Age Structure of the Population as on 1st March

(million)

Year	1997	2002	2007	2012
under	15 353.	64 345.11	334.80	337.93
15-59	532.60	610.55	692.64	758.61
60+	63.64	71.94	84.01	98.50
Dependency ratios :				
(a) Young	0.66	0.57	0.48	0.45
(b) Old	0.12	0.12	0.12	0.13
(c) Total	0.78	0.69	0.60	0.58

Note:

(1) dependency ratio for young is the ratio between under 15 population to 15-59 population

(2) dependency ratio for old is the ratio between 60+ population to 15-59 population

SOURCE: Report of the Technical Group on Population Projections constituted by Planning Commission, published as: Population Projections for India States 1996-2016, Registrar General, India (1996)

Table 1-5 :Population and Labour Force Growth

	Population		Labour Force	
	millions	% growth	millions	% growth
1978	637.6	-	255.8	-
1983	718.2	2.16	286.6	2.07
1994	895.0	2.10	368.5	2.39
1997	951.2	1.85	397.2	2.27
2002	1028.9	1.57	449.6	2.48
2007	1112.9	1.57	507.9	2.44
2012	1196.4	1.45	562.9	2.06

Note: Population and Labour Force figures are based on Tables 4.5 & 4.9

The details of the labour force participation rates and its likely evolution over time are given in Chapter 4. However, the likely position of labour force growth relative to the growth rate in population is given in Table 1-5. It will be seen, that the growth rate of labour force is likely to be highest during the Ninth Plan period and only gradually decline thereafter. As a result, any planning over the perspective period will have to have employment generation as its central focus. If adequate work opportunities are not created, there would be growing incidence of unemployment in the country and an increase in the social tensions that arise therefrom.

Growth Target for Full Employment : Perspective

The need for providing adequate work opportunities for the growing labour force and also to reduce the base level of unemployment significantly has been taken as the most pressing task for determining the minimum required growth performance of the economy. The planning model generates steady state growth paths between the terminal year of the perspective period and the base year in a manner which ensures both inter-sectoral consistency and consistency between the investment requirements and the availability of investible resources. Using constant sectoral incremental capital output ratios (ICORs) and constant sectoral employment elasticities, alternative growth paths can be evaluated in terms of their impact on the generation of work opportunities. The details of the ICORs and the employment elasticities are discussed in Chapters 2 & 4 respectively.

The base line scenario was developed on the basis of a steady state growth path yielding 6.5 per cent average growth per annum over the perspective period. This was consistent with the growth rate that has been observed on the average during the Eighth Plan period. It was found that the work opportunities generated along this growth path were such that the unemployment rate would actually increase from 1.5 per cent in the base year of the Ninth Plan to 2.5 per cent in the terminal year of the perspective period (2011-12). In other words, maintenance of the Eighth Plan growth performance is likely to be inadequate not only to reduce the level of unemployment but also to absorb the additions to the labour force. An alternative growth path targetting an average annual growth rate of 7.4 per cent during the perspective period has also been examined. At this rate of growth the unemployment rate reduces significantly to-3.5 per cent in the terminal year. In other words, not only are the new additions to the work force gainfully absorbed and the backlog of unemployment is removed, but also that the realisation of full employment is likely to occur significantly earlier than the 15 year time horizon. The macro economic parameters of these alternative scenarios are presented in Table 1-6.

Table 1-6 : Alternative Growth Perspectives

(per cent)

	Eighth Plan	15 year	Perspective Plan
		Scenario I	Scenario II
1. Growth rate of GDP	6.8	6.5	7.4
2. Investment rate	24.9	27.4	29.8
(a) Private	16.6	18.5	21.2
(b) Public	8.3	8.9	8.6
3. Saving rate	23.8	25.3	27.2
(a) Private	22.4	23.6	25.1
(b) Public	1.4	1.7	2.1
4. CAD to GDP ratio	1.1	2.1	2.6
5. ICOR (absolute)	3.7	4.2	4.0
6. Unemployment rate	1.5	2.5	-3.5

(Terminal year)

Note :

(1) Growth Rates of GDP is at Factor Cost

(2)Investment and Saving rates are expressed as ratios to GDP at market prices.

(3) Unemployment is on Usual Status basis.

It would be seen that the investment requirement of the two scenarios are not as different as could be expected from the difference in the growth rates. The principal reason for this is that in Scenario-I the ICOR increases sharply from the 3.7 level achieved during the Eighth Plan to 4.2. In Scenario-II, however, the value of ICOR remains at 4.0. This difference arises principally due to the fact that the share of investment in infrastructure, which are the sectors with the highest ICORs, in total investment is higher in Scenario-I than in Scenario-II. This is partly due to the relatively high growth rates that have been assumed for infrastructural sectors in view of the serious shortfalls that have been experienced in these sectors. This is especially true for Rail and Other transport and for Electricity. The reduction in ICOR in Scenario-II does not, however, occur principally by increase in the growth rate of agriculture, which is assumed to rise to 4.8 per cent per annum in Scenario-II as compared to 3.9 per cent in Scenario-I, which is in line with the higher demand arising out of the higher growth rate, but more due to an increase in the shares of manufacturing and services. The broad sectoral growth rates in the two scenarios are presented in Table 1-7.

Table 1-7 : Sectoral Growth Rates in Perspective Plans

(per cent per annum)

Sectors	Scenario I	Scenario II
Agriculture & Allied	3.9	4.8
Manufacturing	8.0	9.6
Infrastructure:		
Group 1	6.3	6.6
Group 2	9.6	10.4
Services	6.8	7.5
GDP at factor cost	6.5	7.4

NOTES : (1) Infrastructure Group 1 contains Construction, Rail and Other Transport.

(2) Infrastructure Group 2 contains Electricity, Gas & Water supply and Communications

1.78 The main difference between the two scenarios in terms of the policy implications lies in the substantial step up that is required in public savings in Scenario-II. The results indicate that unless public savings, particularly the savings of the Government, is enhanced substantially it would not be possible to attain the higher growth path. The alternative to this would be to increase the inflow of foreign savings through a wider CAD. While this may be a viable option, it would depend upon the ability of the country to attract sufficient foreign direct investment (FDI) to finance the foreign exchange requirements without having to resort to increasing external debt or other financial liabilities. In other words, the principal responsibility for generating adequate work opportunities to reduce unemployment in the country and to improve the quality of life rests with the Government, which would need to reduce its existing revenue deficit in the budget and eventually bring it to a surplus. As in all these exercises, the Government includes both the Centre and the States, but the division of responsibility between the two cannot be determined in such an aggregated model. This would depend upon a realistic assessment of the responsibilities of the different wings of Government and their ability to raise the necessary resources. Such an exercise can only be done for a shorter period and would need to be confined to the immediate future. The possibilities that exist during the Ninth Plan period are detailed in Chapters 2 & 3.

Table 1-8 : Planwise Decomposition of Perspective Period

(per cent)

		IXth Plan	Xth Plan	XIth Plan
1.	Growth rate of GDP	6.5	7.7	8.1
2.	Investment rate	28.2	29.8	31.3
(a)	Private	18.8	21.5	23.3
(b)	Public	9.4	8.3	8.0
3.	Saving rate	26.1	27.3	28.2
(a)	Private	24.5	25.3	25.6
(b)	Public	1.6	2.0	2.6
4.	CAD to GDP ratio	2.1	2.6	3.1
5.	ICOR (absolute)	4.3	3.9	3.9
6.	Unemployment rate	1.8	neg.	-3.5

Note :

(1) Growth Rates of GDP is at Factor Cost

(2) Investment and Saving rates are expressed as ratios to GDP at market prices.

(3) Unemployment is on Usual Status basis.

Although the perspective is developed under steady state growth assumptions, it is normally not feasible to change the required parameter values right from the initial years of the perspective period. It is therefore necessary to divide the perspective period into its constituent five year plan periods in order to assess the feasibility of attaining the required parameter values. A possible composition of the 7.4 per cent growth perspective plan is presented in Table 1-8.

As would be seen, the parametric values appear to be quite reasonable, and should be achievable with some effort on the part of the Government and the public sector. The Ninth Plan is the medium-term characterisation of this growth path.

GROWTH PERFORMANCE AND POVERTY

The most significant feature of the 7.4 per cent growth perspective is that over the course of the next 15 years private consumption is likely to grow at marginally above 7 per cent per annum in real terms. With a growth rate of population averaging 1.54 per cent per annum during this period, the per capita private consumption is likely to grow at an average annual rate of 5.5 per cent. In other words, per capita private consumption in India in the terminal year of the perspective period may be over 2.2 times the level that is likely to be achieved in the base year of the Ninth Plan. With such substantial increase in per capita private consumption, the incidence of poverty in the country can potentially reduce significantly.

The Planning Commission has been estimating the incidence of poverty at the national and state levels as an integral component of formulating appropriate growth strategies. For the Ninth Plan, the quantitative index of poverty has been developed on the basis of the recommendations of the Expert Group on Estimation of Proportion and Number of Poor, constituted by the Planning Commission in 1989. The Expert Group recommended computation of state-wise poverty estimates on the basis of state specific poverty lines and the National Sample Survey (NSS) consumption distribution without any adjustment for the discrepancy between the CSO and NSS estimates of private consumption. The estimates of state-wise poverty since 1973-74 have been revised on the basis of the Expert Group methodology. The latest estimates of incidence of poverty are available for the year 1993-94 on the basis of NSS consumer expenditure data of 50th Round (July, 1993 to June,

1994). The poverty ratios and the number of poor at national level for the different years are given in Table 1-9.

Table 1-9 : Percentage and Number of Poor

Year	Poverty Ratio			No. of poor (in million)		
	Rural	Urban	Combined	Rural	Urban	Combined
1973-74	56.4	49.0	54.9	261.3	60.0	321.3
1977-78	53.1	45.2	51.3	264.3	64.6	328.9
1983	45.7	40.8	44.5	252.0	70.9	322.9
1987-88	39.1	38.2	38.9	231.9	75.2	307.1
1993-94	37.3	32.4	36.0	244.0	76.3	320.3

Note : The estimates are based on the methodology recommended by the Expert Group on Estimation of Proportion and Number of Poor as adopted by the Planning Commission

The incidence of poverty expressed as percentage of people below the poverty line is observed to have declined from 56.4 per cent in 1973-74 to 37.3 in 1993-94 in rural areas and from 49.0 per cent in 1973-74 to 32.4 per cent in 1993-94 in urban areas. For the country as a whole, the percentage of people below the poverty line declined from 54.9 per cent in 1973-74 to 36 per cent in 1993-94. However, the number of poor in the country remained more or less stable at around 320 million due to the rise in population. This fact should not over-shadow the 300 million persons who joined the non-poor group during the twenty year period between 1973-74 and 1993-94, but underlines the need for containment of the growth rate of population.

The state-specific poverty lines in rural and urban areas for 1973-74 and 1993-94 are given in Annexure-1.1. The state-wise poverty ratios and number of poor in rural and urban areas as well as for the state as a whole for these two years by Expert Group method are given in Annexures-1.2 and 1.3. It may be seen that the decline in poverty ratio during the last 20 years has been uneven among the states. The pace of poverty reduction was relatively rapid in Kerala, Andhra Pradesh, Tamil Nadu, Gujarat, Punjab and West Bengal. The decline in poverty ratio, however, was not enough to reduce the number of poor in eight major states. These are : Assam, Bihar, Haryana, Himachal Pradesh, Madhya

Pradesh, Maharashtra, Orissa and Uttar Pradesh. Poverty is closely related with hunger. The percentage of persons without two square meals a day for some months of the year is often termed as the "hunger ratio". It is observed that about 19 per cent of the rural population and 7 per cent of the urban population remained hungry in 1983. As per the latest available information these have come down to about 5 per cent in rural areas and 2 per cent in urban in 1993-94. The percentage of chronically hungry people in total population in rural areas, which was 35 to 40 per cent in Bihar, Orissa and West Bengal in 1983, reduced to 6.6 per cent in case of Bihar and 14 to 16 per cent in other two states in 1993-94. Hunger ratios less than the poverty ratio indicate that all the poor people do not remain hungry. However, any incidence of chronic hunger is still a cause of concern.

In order to examine the likely effects of the relatively faster growth of private per capita consumption on the incidence of poverty during the next 15 years, it is necessary to project the behaviour of the poverty ratio on the basis of the trends that are likely to obtain in the macro-economic variables. There are a number of ways in which such projections can be made. One method would be to relate the past trends in poverty ratios with major macro-economic variables such as the growth rate of GDP, the share of agriculture, inflation rate, and policy variables such as public expenditure on anti-poverty programmes, using econometric methods. This approach, however, has the disadvantage that it is unable to capture certain important facets of the behaviour of poverty and inequality in the Indian economy. An alternative methodology is to explicitly recognise these specific features and to extrapolate the future trends on the basis of specific assumptions being made regarding the future movements of these variables which are directly involved in the estimation of poverty. The projections on poverty made for the perspective period are based on this latter approach.

The behaviour of various measures of poverty and inequality in India are given in Table 1-10. The normal method of measuring poverty in India is by the poverty or "headcount" ratio, which is simply the proportion of the number of people below the poverty line in the total population. This ratio, however, makes no distinction within the broad category of the poor depending upon their actual levels of consumption and deprivation. As a result, the poverty ratio fails to capture the depth and severity of poverty in an adequate manner. A measure for capturing the depth of poverty is the poverty gap (PG) index, which adjusts the poverty ratio with the difference between the per capita consumption of the poor and the poverty line expressed as a percentage of the poverty line. This is therefore a measure of the magnitude of the effort that would be required to shift the consumption of all persons below the poverty line to the level of the poverty line. A more comprehensive measure of the severity of poverty is the squared poverty gap (SPG). The SPG is composed of not only the poverty ratio and the poverty gap ratio, but it also captures the consumption distribution of the poor as measured by the coefficient of variation. This measure explicitly takes into account the higher intensity of efforts that are required to address the people who are progressively further below the poverty norm in order to bring them out of poverty.

It will be seen that the annual average rate of decline of the poverty ratio during the period from 1973-74 to 1993-94 has been 2.05 per cent in both rural and urban areas, and 2.09 per cent for the country as a whole. The poverty gap ratio during this period declined by 3.31 per cent per year in rural areas, 2.71 per cent per year in urban areas and 3.21 per cent per year for the country as a whole. The SPG declined even faster by 4.38 per cent, 3.07 per cent and 4.13 per cent per year in rural, urban and the country respectively. The fact that the more distribution-sensitive measures of poverty have declined progressively faster during the last 20 years is a cause for optimism. It implies that the depth and severity of poverty has reduced faster than the incidence, and that over time there has been a trend towards a greater degree of clustering of the poor around the poverty line leading to a situation where economic growth is likely to have a considerably larger impact on poverty reduction than before. This would of course require that the measures of relative inequality do not deteriorate sufficiently to overcome the growth effect.

The most commonly used measure of relative inequality is the Lorenz ratio. The behaviour of the Lorenz ratio of household consumption expenditure as given by the NSS surveys shows a very high degree of stability, as can be seen in Table 1-10. The near constancy of this measure of inequality for 20 years gives a certain degree of justification to believe that it may continue to remain more or less stable in the future as well. However, there is reason to believe that the measures of inequality based on the NSS consumption distribution may actually understate the true level of inequality in the country. A characteristic of consumption behaviour in India, specially in the recent past, has been that the discrepancy between private consumption expenditure as measured by the CSO and household consumption expenditure as obtained from the NSS has been growing. Although a part of this discrepancy may be attributed to consumption by non-household private entities, such as voluntary agencies and non-governmental organisations, the possibility cannot be ruled out that there is a growing incidence of under-reporting of consumption expenditures by the higher income categories.

The methodology for projecting the likely behaviour of the incidence of poverty in the country is based on these two specific features of consumption behaviour in the country namely : (a) stability of the Lorenz ratio for household consumption expenditure; and (b) the increasing discrepancy between the CSO and NSS measures of total non-governmental consumption. Since the plan model yields projections of total private consumption, which is equivalent to the CSO measure, the analogue of the NSS measure of consumption has to be derived by projecting the behaviour of the discrepancy term. This is done by relating the extent of discrepancy to the behaviour of commodity structure of output in the economy over the perspective period as obtained from the inter-sectorally balanced growth path given by the plan model.

The household component of consumption as per the latest estimate of NSS for 1993-94 is 59.7 per cent of total private consumption as given by the CSO and, more importantly, this component for various reasons has steadily reduced in the past 10 to 15 years. For the projections, the household component of the total private consumption has been worked out for three different commodity groups, viz., foodgrain, other food and non-food. The over-all discrepancy between the CSO and NSS estimates of consumption can be decomposed by these three commodity groups, and it is found that in 1993-94 the NSS consumption of food-grains was virtually identical with the CSO estimate, that for other food was 65 per cent of CSO, while that for non-food was only 27.7 per cent. The over-all figure of the NSS analogue of consumption for the perspective period is worked out as a weighted average of these three groups under the assumption that the discrepancy factors remain constant at the 1993-94 values, and with the commodity structure of CSO consumption in the base year as weights. It should be clear from the above estimates of the commodity group-wise discrepancy factors that a relatively higher growth rate of agriculture will lead to a lower increase in the over-all discrepancy term. In other words, this approach explicitly incorporates the observed positive relationship between agricultural growth and poverty reduction by recognising that the trend in increasing implicit inequality has more to do with non-agricultural sectors than with agriculture.

The over-all rates of economic growth of 6.5 per cent per year in the Ninth Plan (1997-2002), 7.7 per cent per year in the Tenth Plan (2002-2007) and 8.1 per cent per year in the Eleventh Plan (2007-2012), as per the selected trajectory, are accompanied by growth rates of per capita household consumption, which is the analogue of the NSS measure of consumption, of 4.3 per cent per year, 4.8 per cent per year and 5.3 per cent per year in the successive plan periods. These are derived from the assumed structure of growth, which yields annual rates of growth of foodgrain, other food and non-food as 3.2 per cent, 5.8 per cent and 7.5 per cent respectively in the Ninth Plan. In the Tenth and Eleventh Plans, the rates of growth of foodgrain and other food remain similar at 3 per cent per year and 6 per cent per year respectively. In case of non-food, the growth rate is assumed to be 8 per cent per year in the Tenth Plan and 8.5 per cent per year in the Eleventh Plan. This structure of consumption growth in conjunction with the group specific household shares for 1993-94 determines the over-all household share of total consumption in the three successive plan periods. The changing pattern of growth in the successive plan periods results in a gradual lowering of the household component. It was 59.7 per cent in 1993-94, and, as a result of the changes in structure of growth, it is projected to go down to 55.6 per cent in Ninth Plan, 53.5 per cent in the Tenth Plan and 50.1 per cent in the Eleventh Plan. As a result, the rates of growth of per capita household consumption are assumed to be significantly lower than the corresponding growth rates of per capita private consumption.

Table 1-11 : Projected Growth in Per Capita Household Consumption In the States

(per cent per year)

Sl. No.	State	Ninth Plan (1997-2002)	Tenth Plan (2002-2007)	Eleventh Plan (2007-2012)
1.	Andhra Pradesh	4.75	5.26	5.83
2.	Assam	4.62	4.81	5.47
3.	Bihar	4.04	4.58	4.99
4.	Gujarat	4.43	4.94	5.52
5.	Haryana	4.38	4.68	5.32
6.	Himachal Pradesh	3.41	4.36	6.00
7.	Jammu & Kashmir	3.53	4.51	6.20
8.	Karnataka	4.61	5.04	5.65
9.	Kerala	5.01	5.31	6.03
10.	Madhya Pradesh	4.06	4.57	5.04
11.	Maharashtra	4.79	5.17	5.80
12.	Orissa	4.99	5.41	6.00
13.	Punjab	4.79	5.05	5.74
14.	Rajasthan	4.08	4.47	4.97
15.	Tamil Nadu	5.05	5.47	6.08
16.	Uttar Pradesh	3.62	4.18	4.47
17.	West Bengal	4.55	5.00	5.59
18.	Other states/UTs	2.39	3.27	4.87
19.	All India	4.29	4.76	5.31

The household component of consumption in the base year of the Ninth Plan and the terminal years of each of the subsequent three plan periods is flowed through a vector of state-wise consumption proportions to allocate the total consumption among the states. This vector of consumption proportions is obtained from the growth in total consumption of each state between 1983-84 and 1993-94 relative to the growth in total national consumption. This procedure implies that the relative position of the states in terms of total consumption witnessed during the period 1983-84 to 1993-94 is assumed to remain unchanged throughout the plan as well as in the perspective period. The underlying assumption is that the rate of growth of consumption in the states during the period are derived from their past performance, and their relative positions do not change. The state-wise growth of per capita consumption in the three future plan periods as derived by this procedure are given in Table 1-11. It should be noted that although the growth in consumption of each state should logically be derived from an assumed growth of state domestic product (SDP), this procedure by-passes this connection and assumes constancy in the consumption shares. No implication can necessarily be drawn about the growth in SDPs from these constant shares, since the state-wise savings behaviour can be quite different from each other.

The state-wise consumption (household component) derived on the basis of the above growth rates are broken up into rural and urban components on the basis of the urban-rural per capita consumption differentials as given by the per capita urban-rural expenditure data of 1993-94, which are assumed to remain unchanged throughout the perspective period, and the population growth in the state. These per capita consumption of the states, which are worked out separately for rural and urban areas, are assumed to follow a two parameter log-normal distribution, which is a standard distribution function used in such analyses and has the distinct advantage that one of the two parameters is directly determined from the Lorenz ratio. As has already been mentioned, the stability of the Lorenz ratio for household consumption expenditure is a critical component of the projection exercise. The dispersion of each of the state-wise distribution functions is computed from the Lorenz ratio of the consumption distribution of each state, using the NSS data on consumer expenditure of 1993-94, and these too are assumed to remain unchanged throughout the three plan periods comprising the perspective.

Table 1-12 : Projection of Statewise Poverty Ratios in the Perspective Period

(per cent)

Sl.No.	State	1996-97	2001-02	2006-07	2011-12
1.	Andhra Pradesh	17.35	11.13	5.35	2.44
2.	Assam	26.46	10.65	3.82	2.07
3.	Bihar	44.09	27.46	14.08	6.52
4.	Gujarat	17.07	9.05	3.94	1.28
5.	Haryana	18.41	10.16	5.00	2.58
6.	Himachal Pradesh	22.38	14.02	7.34	3.14
7.	Karnataka	30.72	17.86	8.68	3.45
8.	Kerala	21.22	11.35	4.76	1.38
9.	Madhya Pradesh	33.38	21.97	12.75	6.81
10.	Maharashtra	32.74	20.66	11.41	5.43
11.	Orissa	40.21	22.93	10.76	4.63
12.	Punjab	8.02	2.98	0.85	0.15
13.	Rajasthan	20.31	10.85	4.70	1.52
14.	Tamil Nadu	30.73	18.11	8.96	3.59
15.	Uttar Pradesh	32.52	21.91	12.88	6.92
16.	West Bengal	25.08	13.54	6.26	2.86
All	India	29.18	17.98	9.53	4.37
All India (weighted)		29.02	17.65	9.19	4.39

From the state-wise distribution functions of consumption expenditure, the state-specific estimates of poverty for the Ninth Plan and for the perspective periods are worked out using state specific poverty lines computed as per the methodology of the Expert Group. These poverty lines have been computed on the basis of the state-wise price indices available for the base year of the Ninth Plan (1996-97) and are assumed to increase at the same rate as the over-all rate of inflation for the projection period. Therefore, this exercise assumes that there is no differential rates of inflation between the states over the perspective period. The poverty ratios are worked out separately for each state in rural and urban areas. The estimate of national level poverty ratio in rural and urban areas have been computed as the weighted average of state-wise poverty ratios as per the Expert Group methodology. In order to check the robustness of the estimates, the incidence of poverty in rural and urban areas at the national level have also been worked out from the national level consumption distribution of the NSS and a national poverty line. The estimates of poverty following this procedure have been worked out for the base and terminal year of the Ninth Plan (1996-97 and 2001-02) and at the end of the Tenth and Eleventh Five Year Plans. The poverty ratios at the national and state levels are given in Table 1-12. As can be seen, the increase in per capita consumption projected over the perspective period is able to reduce substantially the poverty ratio in all the states, including the poorest states.

These estimates of state-wise poverty, however, are critically dependent on three important assumptions : (a) unchanged inequality in the distribution of consumption; (b) unchanged urban-rural per capita consumption differentials; and (c) unchanged poverty norms for all states and the nation. The likely validity of each of these assumptions needs to be examined. The structure of consumption growth in the various states when compared with the pace of poverty reduction reveals that reduction in poverty in all the states is not uniformly dependent on the rise in per capita consumption. Some of the poor states are projected to experience a rate of poverty reduction which is higher than that at the national level despite a lower rate of growth in per capita consumption as compared to the national average. Bihar, and to some extent Orissa, are examples. The projected poverty reductions in most of the poorer states are quite significant in both rural and urban areas. It appears that the states with relatively lower initial inequality in consumption distribution have been projected to experience a faster reduction in poverty than the states with a relatively higher inequality. This phenomenon arises out of the sensitivity of the analysis to the assumed constancy of the Lorenz ratios. It is recognised, however, that as the more backward states start developing, the relative inequality measure may in fact worsen and become progressively similar to those obtaining for the more industrialised states. As a consequence, the reduction in poverty in these states may not be as sharp as projected above. On the other hand, the proportionately higher growth rate in agriculture relative to the growth rate of GDP projected for the perspective period as compared to the historical proportions suggests that the urban-rural consumption differentials may indeed narrow, which would argue for a sharper reduction in poverty than projected. In this context it is interesting to note that an agriculturally developed state like Punjab has lower inequality than the more industrialised Maharashtra. This appears to suggest that the focus that is being placed on agriculture in the Ninth Plan is more likely to be poverty reducing than an industry-based growth strategy. In so far as the constant poverty lines are concerned, it is not possible to project the inflation rates for different states and therefore no statement can be made regarding the impact of differential rates of inflation. Finally, it needs to be mentioned that the observed clustering of the poor around the poverty line and its implication for a faster reduction in the poverty rate has not been taken into account in the above exercise. If this trend continues, the pace of poverty reduction may accelerate even more than projected in this exercise.

The projected poverty ratios in the rural and urban areas at the national level are presented in Table 1-13. The growth in per capita consumption between 1993-94 and 1996-97 reduces the incidence of poverty to 30.55 per cent in rural reas, 25.58 per cent in urban areas and 29.18 per cent for the country as a whole. The incidence of poverty in the terminal year of the Ninth Plan is estimated as 18.61 per cent in rural areas, 16.46 per cent in urban areas and 17.98 per cent for the country as a whole. By the end of the perpsective period the povery ratio is projected to be less than 5 per cent. The poverty reductions are the impact of consumption growth alone as the inequality in the distribution of per capita household consumption is assumed to remain unchanged during the perspective period. Indeed, since the discrepancy between household and private consumption is projected to increase, there is an implicit inference that the extent of relative inequality may rise to a certain extent. Under the assumptions made, the reduction in poverty is higher in rural areas as compared to urban in the Ninth Plan and also in the perspective period. This arises almost entirely due to the fact that relative inequalities are more pronounced in urban areas as compared to the rural.

Table 1-13 : Projection of National Poverty Ratios

(percent)

Region	1996-97	2001-02	2006-07	2011-12
Rural	30.55	18.61	9.64	4.31
Urban	25.58	16.46	9.28	4.49
Total	29.18	17.98	9.53	4.37

1.98 The projections of statewise and national ratios for the perspective period are based on the assumed stability of two key parameters: (a) the Lorenz ratio; and (b) the per capita urban-rural consumption differential.

Table 1-14: Projection of National Poverty Ratios:Sensitivity Analysis

(percent)

Scenario	1996-97	2001-02	2006-07	2011-12
1. Scenario I	27.17	16.00	8.01	3.51
2. Scenario II	28.99	17.84	9.45	4.35
3. Scenario III	26.98	15.86	7.93	3.48
4. Scenario IV	25.08	14.02	6.60	2.79
5. Scenario V	28.81	17.71	9.37	4.32
6. Scenario VI	24.69	13.76	6.48	2.75
Base Case	29.18	17.98	9.53	4.37

Notes:

1.Scenario I is based on a reduction of Lorenz ratio by 5%, i.e., from 0.2816 to 0.2675 in rural areas and from 0.34 to 0.323 in urban areas.

2.Scenario II is based on a reduction of per capita urban-rural consumption differential by 5%, i.e., from 62.78% to 59.64%.

3.Scenario III is based on the impact of both Scenario I and Scenario II.

4.Scenario IV is based on a reduction of Lorenz ratio by 10%, i.e., from 0.2816 to 0.2534 in rural areas and from 0.34 to 0.306 in urban areas.

5.Scenario V is based on a reduction of per capita urban-rural consumption differential by 10%, i.e., from 62.78%to 56.50%.

6 Scenario VI is based on the impact of both Scenario IV and Scenario V.

The rationale for the assumed constancy of these ratios lies in their past behaviour. However, public policy and development strategy can conceptually affect their future behaviour. In order to understand the magnitude of such effects, the sensitivity of the poverty measures to changes in the distributional parameters need to be seen. The impact of a reduction in inequality on poverty worked out through a simulation exercise shows that 5 per cent reduction in the Lorenz ratio of consumption distribution, *ceteris paribus,* reduces the incidence of poverty by an additional 2 percentage points in the terminal year of the Ninth Plan and reduction in the Lorenz ratio by 10 per cent reduces the incidence of poverty in the Ninth Plan by 4 percentage points. On the other hand, reduction in the urban-rural per capita consumption differential by 5 and 10 per cent reduce the poverty ratio by only 0.2 and 0.4 percentage points respectively. The principal reason for this limited effect is that while rural poverty declines, there is a partial off-set through increase in the urban poverty. For this reason, it may not be desirable to focus excessively on policies to reduce the urban-rural differential except on the basis of distribution of the incremental growth. The national poverty ratios arising from the sensitivity analysis are presented in Table 1-14.

There are a number of ways by which the reduction in inequality in consumption distribution can be effected. The increased outlay on anti-poverty programmes, if directed towards states with higher poverty and high poverty regions within the states, can lead to an improvement in the inequality parameters and thereby to a faster reduction in poverty than projected. Another option lies in transferring consumption from top to the bottom deciles of the population. A 5 per cent rise in the share of consumption of bottom three deciles of the population at the expense of top 10 per cent means a reduction in the income of the top 10 per cent by 2 to 3 per cent of their consumption share. In a similar way, a 10 per cent rise in the share of consumption of bottom three deciles of the population means a reduction in the income of top 10 per cent of the population by 5 to 6 per cent of their consumption share. It needs to be realised, however, that although the percentage changes in the distributional parameters appear to be fairly small, the magnitudes involved are quite substantial and may involve considerable effort. For example, a 5 per cent and 10 per cent rise in the share of consumption of bottom 30 per cent of the population involve a transfer of 0.75 per cent and 1.5 per cent of the total consumption respectively. This translates into an annual transfer of total income of Rs. 4753 crore and Rs. 9505 crore at 1996-97 prices during the course of the Ninth Plan for the two cases. These transfers are in addition to the existing redistributive measures through anti-poverty and other programmes which are implicitly built into the Ninth Plan. The effect of these transfers would be to reduce the incidence of poverty from 18 per cent in the terminal year of the Ninth Plan in the base-line to 15.9 per cent and 13.8 per cent respectively. It needs to be understood that the income transfer from the top to bottom deciles may not be reflected entirely in reduction of the consumption of the former. A certain portion of the reduction in income would almost inevitably fall on their savings. Thus, such a transfer is likely to result in a reduction of the savings rate in the economy. The magnitude of this shortfall, in itself, may not be a serious problem, provided it is made up through increased efforts towards mobilising additional investible resources.

These projections of the possible reduction of poverty during the perspective period, though quite impressive, should not be a cause for complacency. In particular, there is no cause to bring into question the utility or size of the anti-poverty programmes at the present time. These programmes, it should be remembered, not only alleviate the immediate deprivation of the poor, but also contribute quite significantly to the creation of productive assets-private assets in the case of schemes like IRDP, and community assets through JRY and EAS. Therefore, the impact of these schemes are embedded in the above analysis both through their direct effect on the consumption of the poor and through the indirect effects arising from the income generated from the assets created. Reduction of these schemes is likely to alter the parametric basis of the analysis and there is no estimate of the magnitude of such changes. More importantly, time series analyses indicate that the existence of these schemes, along with the PDS, have prevented the distribution-sensitive measures of poverty, such as the PG and SPG indices, from worsening during natural calamities. Closure of these programmes would expose the population below or around the poverty line to vulnerabilities which continue to be present in the Indian economy. It should be recognised that starting new programmes during times of necessity is expensive in terms of both human and financial costs and also tends to be inefficient. There is, however, a strong case for better targetting and improving the effectiveness of these programmes.

ENERGY AND NATURAL RESOURCES

Achievement of an average 7.4 per cent growth rate in GDP over the next 15 years will generate a high growth in demand for natural resources, particularly energy. The demand for total primary commercial energy increased from 25.5 MTOE in 1953-54 to 212.9 MTOE in 1994-95, representing an increase of more than 8 times over the 40 year period. It is likely to increase to 770 MTOE by 2011-12 representing over 3 fold increase in 17 years. In per capita terms, the demand for primary commercial energy was 67 KGOE in 1953-54. It increased to 234 KGOE in 1994-95. By 2011-12 the per capita primary commercial energy consumption is expected to increase to 648 KGOE. Despite the not inconsiderable growth, this level of per capita primary commercial energy consumption in 2011-12 is low when compared to the international average per capita energy consumption levels of 1433 KGOE in 1994-95, and over 5000 KGOE in the developed countries. The demand for electricity, in particular, is likely to increase by about 3.5 times by 2011-12 as compared to 1996-97. As a result, more installed generating capacity will have to be added over the next 15 years than what has been done in the last 50 years. The demand for petroleum products is likely to treble by 2011-12 than what it was in 1996-97. Similarly, the demand for coal is likely to increase by nearly 2.6 times. These will require increase in exploration activity, development of mines/oil fields, setting up of thermal and hydro-electric power stations which are not only highly capital intensive but also are likely to put great pressure on the quality of environment. Such large demand will also require transport facilities viz. railways, roads and ports to be expanded sufficiently so that these infrastructural facilities do not act as a constraint for achieving the desired growth in the economy in general and the availability of energy in particular.

In the long run, some of the important questions that the planners will have to themselves address to are: (a) how long will depletable resources, such as oil, last; (b) what sort of export strategy will have to be adopted so that growing requirement of imported energy is met without adversely affecting the balance of payments position; (c) what are the environmental consequences associated with intensive mining and use of the relatively abundant mineral resources such as coal, iron-ore, etc. In short, the question of sustainability of natural resources and the environment as well as that of strategic self-reliance are important in the long run and will need to inform the development strategy. This question of sustainability has to first take account of the domestic availability of these resources and the options that are available to the country. The domestic availability and the life indices of some of the important mineral resources is given in Table 1-18. Even though the size of certain primary commercial energy reserves appear fairly large, their per capita availability is modest on account of the large population of the country. If no significant additions are made in future, the per capita availability would worsen.

As can be observed from the table, the hydrocarbon reserves of the country are unlikely to last beyond 2016 (oil reserves, in fact, will last only till 2014) at the projected production levels of 2001-02, and that of copper beyond 2009. The position regarding other minerals is comparatively better. However, mining is beset with environmental problems as it often leads to deforestation, diversion of land from agricultural to other purposes, etc. Greater use of energy would result in increased carbon emission and other pollutants. This coupled with denudation of forests (which act as sinks for emissions) is likely to worsen the environmental problems.

The demand for electricity (generation requirement) is likely to increase to 1473 Bkwh by the end of the Eleventh Plan, as against 394.5 Bkwh in the utilities in 1996-97 representing an average increase of 9.1% per annum. Much of the generation is likely to be thermal, through within thermal there is a possibility of greater share of hydro-carbon (oil & gas) based generation vis-a-vis coal based generation. Greater emphasis on hydro-electricity is important, particularly to meet the peak requirement. Hydro-electricity should also be preferred as it is a clean source of energy. However, much of the hydro-electric potential that remains to be tapped is located in the ecologically fragile Himalayan region and in more difficult locations. Tapping this source, therefore, is fraught not only with environmental hazards but is also much costlier as compared to thermal power. On the other hand, coal based thermal power plants lead to problems related to air pollution, ash disposal, etc. Dependence on oil based thermal power plants would result in less pollution, less pressure on railways and other infrastructure, but it would result in greater dependence on imports and the consequent reduction in self-reliance, and also increase balance of payments pressures. An important source of energy which has been neglected over the recent years is nuclear power. Although it is relatively expensive in terms of the capital costs, it is free from many of the problems discussed above. Consideration needs to be given to increasing the share of nuclear power as a viable alternative source of energy.

The rate of growth of nuclear power is currently limited by financial constraints while it is technologically poised for higher rates of growth. The availability of nuclear fuels in the country can sustain 350,000 MWe of power, and the industrial capability and R&D infrastructure can indeed support an accelerated programme. The nuclear power programme should grow to a minimum size of 20,000 MWe in the course of the next 25 years. The programme would be largely based on indigenous designs of Pressurised Heavy Water Reactors and Fast Breeder Reactors. Some Light Water Reactors may be inducted as a supplementary line to increase the nuclear power capacity at a faster pace, subject to such systems being available on political and commercial conditions acceptable to the country. A reasonable size Nuclear Power Programme based on fission systems will also enable the country to develop fusion based systems (using hydrogen as nuclear fuel) in due course of time. Such fusion based systems are expected to play a very important role in the long run in meeting the energy needs of mankind. Full benefits of this technology will not be available to the country unless its growth is adequately supported in the near term as a matter of policy. Nuclear power development is important to avoid excessive dependence on energy import, in the coming decades.

The major problem of the power sector is the optimum generation mix. In the short run, dependence on gas based and oil based plants seems to inevitable for meeting the power demand on account of the relatively shorter gestation periods of these sources. However, in the long run, the optimum mix has to be planned in such a manner that bulk of the base load requirements will have to be met from coal based thermal electricity, supplemented by nuclear electricity to the extent possible, while the peak requirement has to be met from hydro-electric stations and oil/gas based power. The use of oil for meeting base load requirements will have to be discouraged. The signing of PPAs in future should ensure that oil/gas fired stations are mainly configured for meeting peaking requirements. These fuels should not be used for meeting the base load requirements. Indeed, there is a case for converting existing oil/gas based power plants and those which are in the process of being established to peak-load plants at some point in time in the future. The agreements with the independent power producers may have to contain clauses which would permit such reassignments.

Sufficient reserves of coal are available in the country. The demand for coal is likely to increase from 296 million tonnes in 1996-97 to 776 million tonnes in 2011-12. The domestic production is likely to increase to 672 million tonnes in 2011-12 as against 286 million tonnes in 1996-97. Thus, the gap between demand and domestic production of coal is likely to be 104 million tonnes in 2011-12. Much of the coal mining in India is done by open cast mining techniques. Although it raises the productivity and recovery of coal from the mines, it leads to severe environmental problems viz. land degradation (including that of agricultural and forests), displacement of people, air pollution, etc. As per some studies done in the Planning Commission, the land requirement for open cast mining was about 513 hectares per million tonnes of coal, as compared to 235 hectares per million tonnes of coal by underground mining. According to the Working Group of Ninth Plan, nearly 790 hectares of land will be required per million tonne of additional coal production. On the other hand, the recovery and productivity in underground mining is lower, which increases the unit cost of production of coal. In the long run, new ways of mining coal, viz. coal slurry or coal bed methane, etc. are some of the technologies that will have to be adopted to reduce the environmental hazards as well as to improve the productivity. Greater dependence on imports of coal would also require massive expansion of port facilities, rail transport network (unless the end-use industries are set up close to ports), apart from putting pressure on the balance of payments. The utilisation rate of coal in all the end-use industries will have to improve in order to reduce the demand of coal and problems associated with its production and use.

The recoverable reserves of crude oil in the country are likely to be of the order of 513 million tonnes by 2001-02if no further accretion to reserves takes place. The production of crude oil is expected to be 37 million tonnes. With hardly any accretion in the reserves taking place in the last few years (in fact, there is a net deceleration in the reserves), the reserves to production ratio is likely to decline. The accretion to crude oil reserves was negative in each of the first three years of the Eighth Plan while in case of natural gas, the recoverable reserves declined from 735 billion cubic metres from 1992-93 to 707 billion cubic metres in 1994-95. Given a constant production level of 37 million tonnes from 2001-02 and the anticipated recoverable reserves of 513 million tonnes, the crude oil reserves are likely to last only till 2015-16. These reserves could be completely exhausted in 2011-12, if it is assumed that 30% of the demand of petroleum products will be met from domestic crude oil production. The situation for natural gas is marginally better. In addition, there also exist potential of coal bed methane, oil shale and methane gas hydrates in the country. As per the initial estimates available, the potential is placed at 850 billion cubic metres of coal bed methane, 600 million tonnes of oil shale and 6156 trillion cubic metres of gas hydrates. If these estimates are confirmed by detailed exploration, there will be a big improvement with regard to availability of energy sources for the growth of our country. It is thus, abundantly clear that until and unless these new hydrocarbon sources are proved and developed, the demand for petroleum products in the years to come will have to be increasingly met from imports. This will put pressure on the balance of payments. Therefore, exports will have to increase substantially. Efforts will have to be simultaneously made to reduce the oil intensity of the economy. This is particularly true of the transport sector. The share of railways in the total freight movement in the last few years has declined sharply vis-a-vis the roads sector. It is important to reverse this trend. The energy efficiency of movement by rail is better than of roads and steps will have to be taken to encourage it. Since the early 1980s, growing number of scooters (2-wheelers) and cars have been plying on the roads, particularly in the urban areas. This has resulted in greater energy (in particular oil) intensity, traffic jams and environmental hazards viz. air pollution. If unchecked, this problem could take serious dimensions in the years to come. It is, therefore, essential that these modes of transport are replaced by public modes of tranport, viz. buses and whenever feasible by suburban trains; including mass rapid transport system (MRTS).

So far as non-conventional energy sources are concerned, the development of wind generators in recent years has been quite impressive. Suitable policy measures to encourage the use of wind, solar, biomass and co-generation should be evolved. Consideration may be given to making use of solar water heating systems mandatory in residential buildings above a certain size. There is need to consider enacting a legislation to enforce energy conservation standards. It is important that conservation of natural resources receives adequate priority, especially from environmental angle. The danger of further deterioration in the quality of air and water is not unreal and so is the case with the quality of soil. Greater demand for land resources for food, fodder, fuel, water as well as for mining and other activities will put pressure on the availability and quality of land. Therefore, it is essential that these resources are used with utmost care so that growth is sustainable.

One of the main problems in taking a rational decision on natural resource use is the lack of an appropriate information system and a methodology for natural resource accounting. As a result, the depletion of the national asset base is simply not taken into account while evaluating alternative strategies. The Ninth Plan will lay great stress on evolving such natural resource accounting methodologies so that decisions can be taken on the basis of the full cost to the Nation. The most important methodology for preventing undue depletion of natural resources is by ensuring their efficient use at all levels. This has two important dimensions. First, exhaustible or vulnerable natural resources should be priced appropriately in order to prevent over-exploitation. In India, since a large proportion of natural resources are subjected to administered prices, there has been a tendency to under-price resources quite significantly. This approach needs to be changed urgently. Second, technologies which conserve the use of natural resources need to be developed and promoted vigorously.

Technology

In order to strengthen the technological capabilities of the Indian industries, both for meeting the national needs and for providing global competitiveness, a number of new initiatives have been launched. A Technology Development Board has been established in 1996 with a mandate to facilitate development of new technologies and assimilation and adaptation of imported technologies by providing catalytic support to industries and R&D institutions to work in partnership.

Matching grants to R&D institutions showing commercial earnings through technology services was also introduced in 1996. Such initiatives will need to be carried forward and broadened. Issues of natural resource conservation and agricultural growth cannot be effectively tackled in the absence of an appropriate technological base. In addition, technology is essential for increasing the competitiveness of the Indian economy in international markets. Indigenous development of technology is therefore of the highest importance and deliberate planned steps need to be taken to increase technological self-sufficiency of the Nation.

Rapid technical progress is altering fundamentally the skills, knowledge, infrastructure and institutions needed for the efficient production and delivery of goods and services. So broad and far-reaching are current technological developments that many see the emergence of another industrial revolution driven by a new technological "paradigm". This paradigm involves, not only new technologies and skills in the traditional sense, but also different work methods, management techniques and organisational relations within firms. As new transport and communications technologies shrink international 'economic space', it also implies a significant reordering of comparative advantage, and trade and investment relations, between countries.

In India also, there is considerable technological activity in a wide spectrum of firms. What is most impressive is the number of small and medium sized enterprises that are investing in new technology based ventures, and often striking out in world market as exporters. However, the rest of the industrial sector still needs to invest on technology upgrading. Experience of many developing and industrialised countries suggests that a rapid acceleration of industrial technology development calls for a deliberate 'strategy', in the sense that it requires the government to coordinate and guide an essentially market-driven process. Free markets suffer from various kinds of 'market failures', they may not throw up the appropriate amounts of infrastructure, skill, information and institutional support, and mere exposure to market forces, while getting rid of inefficient policies, may not suffice to create the technological dynamism that continued industrial growth needs.

Indian technology policies are undergoing significant changes, and on the whole have improved greatly in recent years. They are not, however, ideal. A coherent technology strategy in India must address a number of interconnected elements in the incentive regime and the relevant factor markets and institutions. The following approach should guide future policies in technology:

Skills : Technology development calls for both general and specific forms of human capital, and emerging technologies are highly skill intensive in both technical and managerial terms. While India is endowed at present with large amounts of high-level human capital, investments in the creation of new skills (as measured by enrolment levels in technical subjects at all levels) are low. In addition, firm level investments in training are highly variable, and large parts of industry invest very little in training. The SME sector in particular suffers from very low levels of skill, while industrial training institutes are often unresponsive to their needs.

Technological effort : R&D in Indian industry has been rising, but the overall level is still low and over three-quarters of research effort originates in the public sector. This is in contrast to Taiwan and Korea, where most of R&D is undertaken by industrial enterprises. It is important for the Government to analyse current technological trends in industry in order to formulate appropriate policies to encourage R&D.

Technology access **:** Technology upgrading requires that Indian enterprises of all types have information on relevant technologies in international markets and also within the country. Many countries have well-developed systems of computerised on-line technology information and dissemination services, often backed up with consultancy and financial assistance for small and medium enterprises to enable them to know about, test, and implement new technologies. The availability of similar facilities are weak in India. Note needs to be taken of the emerging trends of restrictions on access to technology through stringent intellectual property restrictions and on "dual use" technologies by certain groups of industrially advanced countries. Special efforts have to be made both at national as well as at international levels, to ensure that such restrictions do not inhibit progress in high technology sectors.

Technology institutions **:** India has a large infrastructure of technology support institutions, some of which are undergoing reform to make them more relevant to industrial needs. A number of universities, especially the IITs, are interacting increasingly with industry on technological matters, while others are outside this circle. There is a need to strengthen 'Technology Foresight Programmes' to analyse the implications of emerging technologies, analyse domestic strengths and weaknesses and target future technologies for local development.

Other infrastructure **:** Technology development generally requires the setting up of clusters of industries that can share information and skills, as in 'science parks' or dedicated industrial estates. Some such facilities exist in India, but their efficacy and functioning need to be strengthened.

GROWTH AND INVESTMENT TARGETS

Accelerating the growth rate of the economy with stable prices is central to the attainment of a number of objectives discussed in the previous chapter. There is by now enough evidence to show that rapid growth has strong poverty reducing effects especially when it is complemented by a public policy stance which is sensitive to the needs of the poor. Accelerated growth will also help in realising the objectives of increasing employment and, through that, spread of income generation and poverty eradication. However, the linkage between growth, employment and poverty reduction depends crucially upon the sectoral pattern of growth and on the degree to which the disadvantaged segments of the population and the backward regions of the country are successfully integrated into the wider growth process.

The growth objective also subsumes a number of subsidiary objectives which have, at one time or another, been explicitly identified as objectives in earlier Five Year Plans. Industrialisation, productivity growth, infrastructural development, are examples of such subsidiary objectives. The logic of economic growth and the requirements of structural consistency demand that most of these subsidiary objectives are met in order to attain the primary objective of accelerating the over-all growth rate. It is therefore usually unnecessary to spell out each component of growth as a distinct objective unless it has direct bearing on the implementation of the over-all growth objective as possible alternative paths of development. It is in this context that agricultural growth has been specifically given the highest priority, as it determines a pattern of development that accelerates GDP growth with a rapid reduction in unemployment and poverty.

Economic growth is the outcome of numerous factors interacting with each other. For resource-constrained developing countries, capital accumulation or investment is the most important factor for increasing the productive capacity of the economy as well as for improving the productivity of the other factors of production. The Indian planning methodology has therefore traditionally focused on the relationship linking growth to the investment rate and the incremental capital-output ratio (ICOR), and the Indian Plans have been essentially investment plans, dealing with the allocation of investible resources among different sectors, maintaining inter-sectoral consistency towards attaining the targetted rates of growth. In a broad sense the results of such plans have not been unsatisfactory. As can be seen in Table 2-1, except for the Third and the Fourth Plans, the actual performance of the economy has usually been at or above the planned or targetted level. During these two Plans, the shortfalls were largely due to exogenous shocks that could not possibly be predicted. The Third Plan witnessed the drought years of 1965 and 1966, and the Indo-

Pakistan War of 1965. The Fourth Plan experienced three consecutive years of drought (1971-1973) and the first oil-price shock of 1973.

Table 2-1 : Growth Performance in the Five Year Plans

(per cent per annum)

		Target	Actual
1.	First Plan (1951-56)	2.1	3.61
2.	Second Plan (1956-61)	4.5	4.27
3.	Third Plan (1961-66)	5.6	2.84
4.	Fourth Plan (1969-74)	5.7	3.30
5.	Fifth Plan (1974-79)	4.4	4.80
6.	Sixth Plan (1980-85)	5.2	5.66
7.	Seventh Plan (1985-90)	5.0	6.01
8.	Eighth Plan (1992-97)	5.6	6.78

Notes :

(1) The growth targets for the first three Plans were set with respect to National Income. In the Fourth Plan it was Net Domestic Product. In all the Plans thereafter, Gross Domestic Product has been used

(2) The Eighth Plan actual is based on the Quick Estimate for 1996-97.

Even today this approach remains useful as a starting point for assessing the limits of the possible alternative paths of development and the steps required for accelerating the growth rate of the economy. However, with the economic reforms of the last few years, some of the traditional instruments of implementing the Plans are no longer available. In particular, for determining the over-all rate of investment or its sectoral distribution, market forces, relative prices and incentives now play much more important roles than direct allocation of resources by public authorities. With the steady reduction in the share of public investment, both planned and actual, in total investment, as shown in Table 2-2, the ability of the Government to determine the structure of the economy through its own investment behaviour has eroded significantly. With the greater importance of private investment and the movement towards a more market based system, planning has to move away from direct intervention strategies to planning for policies. These issues will be addressed in this chapter.

Table 2-2 : Share of Public Sector in Total Investment

(per cent)

	Planned	Realised
Fifth Plan (1974-79)	57.6	43.3
Sixth Plan (1980-85)	52.9	47.8
Seventh Plan (1985-90)	47.8	45.7
Eighth Plan (1992-97)	45.2	34.3

The incremental capital-output ratio (ICOR) is a summary expression for the existing technical conditions and structural configuration of the economy which captures the relationship between investment and additional output. The ICOR is usually expressed as a relationship between investment and output in the same Plan period. However, there are gestation lags which are present in any investment activity, and the ICOR depends in fact on the pace of investment in the past creating additional output during the Plan period and also takes into account the investment which will create capacities for sustained growth in the future. In addition, different sectors of the economy have very different ICORs, and the aggregate ICOR for the whole economy is a weighted average of these sectoral ICORs. It depends crucially on the sectoral composition of investment and growth. The sectoral structure of growth, in turn, depends on a number of factors such as the pattern of demand, the nature of inter-sectoral linkages, and the possibilities of trade. Attempts are made to capture these complexities by using an input-output matrix based planning model. It needs to be kept in mind that the aggregate ICOR which has been used for discursive purposes in formulating this Plan is derived from a model which tries to capture the gamut of complexities which are inherent in any economy.

Since its inception, the planning methodology in India has been based on the assumption that the economy is savings constrained. In other words, it is assumed that the demand for investible resources always exceeds the supply, and hence the total level of investment is determined uniquely by the available level of savings. While this assumption was more true in the past than at present, with increased opportunities of borrowing from abroad and of foreign savings supplementing domestic savings, it is true that investment cannot exceed the total availability of savings for any length of time without incurring the risk of unsustainability. Indeed, the availability of total savings over a period determines the maximum sustainable level of investment, and hence of growth.

The total volume of investible resources or total savings available in the economy is determined both by the level of domestic savings and by the inflow of foreign savings financing a desirable level of current account deficit. Domestic savings originate from three principal sectors, namely: (a) the Government, including the public sector; (b) the private corporate sector; and (c) households. The inflow of real foreign savings is the current account deficit and can be financed either in the form of debt, both public and private, or foreign investment, both direct and portfolio.

The growth prospects for the future need to be based on an analysis of the recent growth experience, a careful assessment of the resources which could be mobilised and the imperatives of meeting certain minimum social needs and aspirations. Care also has to be taken to ensure that the growth process is sustainable both in terms of external indebtedness of the country and the fiscal viability of the Government not only in the current Plan period but in the future as well.

Since the Plan has commenced from April 1, 1997 and cover the five-year period from 1997-98 to 2001-02, the base year for the exercises is 1996-97. All the calculations are presented at prices that are assumed to prevail in 1996-97.

The average growth rate that has been achieved in the Eighth Plan in terms of the GDP at market prices is 6.8 per cent. This has been supported by an average investment rate of 25 per cent of GDP, thereby yielding an ICOR of 3.7, which is significantly less than the figure of 4.1 that was assumed in the Eighth Plan calculations.

There has been a significant step-up in the domestic savings rate during the Eighth Plan period to 24 per cent of GDP as compared not only to the long period of stagnation experienced during the 1980s, but also in relation to the savings target assumed for the Eighth Plan, which was placed at 21.6 per cent. This is a positive development and augurs well for the future. On the other hand, the reduction in the over-all ICOR should have been sharper in view of the perceptible improvement in the growth rate, particularly in agriculture, and hides certain adverse developments. The principal cause of concern relates to significant increases that have taken place in the ICORs of some of the infrastructural sectors.

A more serious problem, however, is that because of fiscal difficulties, total public investment has slipped quite sharply from the Eighth Plan targets. During Plan formulation it had been assumed that public investment would form about 45 per cent of total investments in the country.

In actuality, however, it is unlikely to be greater than 34 per cent, dipping to below 30 per cent in the closing years of the Plan as shown in Table 2-3. On the other hand, total investment at 25 per cent of GDP has significantly exceeded the Eighth Plan target of 23 per cent. Taking this into account, the slippage in public investment as a percentage of GDP has been from the target of 10.4 per cent to 8.3 per cent. This slippage of 2 percentage points has fallen disproportionately on infrastructure, both economic and social. On the positive side, there appears to be some buoyancy in private investment which has increased from the target of 12.6 percent of GDP to 16.6 per cent.

Table 2-3 : Investment Shares in the Eighth Plan

	92-93	93-94	94-95	95-96	96-97	Plan
1. Total Investment (% of GDP)	22.4	22.7	25.6	27.1	27.3	25.0
2. Public Investment (% of Total Investment)	36.2	38.8	35.5	29.0	29.3	34.3
3. Public Investment (% of GDP)	8.1	8.8	9.1	7.9.	7.4	8.3

Note: These shares are at current prices

Keeping these factors in mind, the macro-economic performance of the economy, that is likely to occur during the Ninth Five Year Plan, had been projected on the basis of a model that had been developed specifically for this purpose. In calculating the parameters of the model for the Ninth Plan period, a number of specific assumptions had been made which need to be noted. First, it had been assumed that the trend towards improvement in total domestic savings, as well as in public savings, witnessed in the last couple of years would continue during the Ninth Plan.

The improvement in Government savings would have to arise essentially from combined improvement in taxes net of subsidies, with public consumption expenditure more or less maintaining its share of GDP.

This assumption explicitly takes into account the impact arising from the adoption of the Fifth Pay Commission recommendations both by the Centre and the States. As a consequence, although public consumption increases sharply during the first two years of the Plan, it reduces equally dramatically in the last three years.

Total public investment, which includes investment by the Centre, the States and all public sector enterprises, had been determined with a view to achieving a fiscal deficit target of 4.0 per cent for the Centre during the Plan period. In view of the emphasis that is being placed on investment in infrastructure, the ICOR would tend to rise from 3.7 during the Eighth Plan to 4.3 during the Ninth Plan. This parameter was generated by the model in order to ensure that the growth rate of the economy would not suffer in the post-Plan period due to infrastructural bottlenecks which could arise from the shortfalls in pipe-line investment that occurred during the Eighth Plan period.

On the basis of the parameters described above, the likely growth rate of GDP during the Ninth Plan in the base-line scenario following the recent trends was projected to average about 6.2 per cent per annum, despite a step-up in both private and public savings, and a current account deficit of about 1.7 per cent of GDP.

Public investment as a whole would be about 33 per cent of total investment in the economy, which would be considerably lower than 45.2 per cent that was assumed during the Eighth Plan and about the same as 34.3 per cent that is likely to be attained. The standard of living of the people, as determined by the growth in private consumption, would be likely to grow by 5.7 per cent per annum, which amounts to a growth in per capita consumption of 4.0 per cent assuming a growth rate of population of 1.7 per cent over the Plan period.

This compares very favourably with the 2.9 per cent and 3.5 per cent annual growth in per capita consumption that were recorded during the Seventh and the Eighth Plan periods, and arises from the lower rate of population growth on the one hand and greater availability of public and foreign savings for financing the required investment, on the other.

On the negative side, the pattern of growth and the likely employment elasticities are such that the employment potential created in the base-line scenario would barely absorb the additions to the labour force during the Plan period. However, because of the increase in infrastructure investment, the conditions are created for an acceleration in the growth rate and consequently a sharp reduction in the unemployment rate during the post-Plan period.

An alternative growth scenario envisaging an accelerated growth rate of 7 per cent during the Ninth Plan was also examined. A preliminary assessment of the macro-economic and sectoral implications and the necessary policy measures which would need to be taken by both the Centre and State governments for attaining such a growth rate were worked out and presented to the National Development Council (NDC).

In January 1997, the NDC approved a GDP growth target of 7 per cent per annum for the Ninth Plan. Detailed exercises were carried out to ascertain the feasibility of this growth target and the manner in which it may be attained. It was found that the growth target is feasible, provided that there is a shift in the development emphasis and certain policy decisions are taken.

The configuration of the macro-economic parameters which would make the target achievable are presented in Table 2-4. Subsequent events, however, have raised doubts regarding the attainment of the growth target mandated by the NDC for the Ninth Plan period. The first year, i.e., 1997-98 of the Ninth Five Year Plan (1997-2002) is already over. The performance of the economy has not been encouraging during this year. According to the latest data available for the year 1997-98, the Gross Domestic Product at factor cost (GDPfc) has grown at a rate of about 5.25 per cent, with the agricultural sector showing a negative growth rate. Due to a variety of reasons, although the economy is expected to pick up momentum during the second year of the Plan, it is unlikely to entirely recover the ground lost in the first year.

Table 2-4 : Macro Parameters for the Ninth Plan (1997-2002)

[for a GDP growth rate of 7 % per annum]

	VIII Plan	IX Plan	Post Plan
1. Domestic Savings Rate (% of GDP at market price)	24.1	26.2	27.2
2. Current Account Deficit (% of GDP at market price)	0.9	2.1	2.4
3. Investment Rate (% of GDP at market price)	25.0	28.3	29.5
4. ICOR	3.9	4.0	3.9
5. GDP Growth Rate (% Per Annum)	6.5	7.0	7.5
6. Export Growth Rate (% Per Annum)	10.3	14.5	14.5
7. Import Growth Rate (% Per Annum)	14.1	12.2	13.2

As a result, the growth trajectory is likely to be adversely affected for the entire Plan period since the absolute level of investible resources will stand revised downwards. In addition, the ICOR is also likely to increase significantly due to the temporary reduction in the level of capacity utilisation. Given these considerations, the growth target of the Ninth Five Year Plan has been revised downward to 6.5 per cent per annum on an average. Although this may appear to be a deviation from the NDC approval, it needs to be noted that the average growth rate of GDP projected for the last three years of the Plan is about 7 per cent. A higher target for this period may not be feasible. In order to attain the growth target for the perspective period, suitable adjustments have been incorporated into the post-Plan scenario. The revised macro parameters consistent with this revised target growth rate are presented in Table 2-4(a).

It can be seen that the strategy that has been selected for attaining an accelerated growth rate of 6.5 per cent hinges on a fairly sharp increase in the investment rate of about 3.3 percentage points above the Eighth Plan average, while keeping the aggregate ICOR at a level somewhat higher than in the Eighth Plan. These are deliberate choices which have been made from a range of possible alternatives on the basis of a number of considerations which are discussed later.

Table 2-4(a) : Macro Parameters for the Ninth Plan (1997-2002)

	VIII Plan	IX Plan	Post Plan
1. Domestic Savings Rate (% of GDP at market price)	23.8	26.1	27.2
2. Current Account Deficit (% of GDP at market price)	1.1	2.1	2.6
3. Investment Rate (% of GDP at market price)	24.9	28.2	29.8
4. ICOR	3.7	4.3	3.9
5. GDP Growth Rate (% Per Annum)	6.8	6.5	7.7
6. Export Growth Rate (% Per Annum)	11.9	11.8	14.5
7. Import Growth Rate (% Per Annum)	11.7	10.8	15.9

The increase in investible resources will have to come mainly from domestic sources, 2.3 percentage points of the required 3.3, and the remainder of about 1.0 percentage points from external savings. Although the ICOR target appears to be a relaxation of the level attained in the Eighth Plan, it has to be seen in the context of the projected increase in the ICOR to 4.3 during the Ninth Plan period in the base-line scenario which, as has been mentioned, would be necessary to create sufficient pipe-line investment in infrastructure for sustaining growth in the post-Plan period. A decline of even this magnitude is by no means trivial. Both these parametric requirements for achieving the target growth rate, therefore, are quite formidable and cannot be assumed to be attained without deliberate planning and strategic policy initiatives.

The macro-economic aggregates arising from the Ninth Plan growth target are presented in Table 2-5. As can be seen, the size of national investment in this scenario rises substantially from Rs. 1399 thousand crore during the Eighth Plan to Rs. 2171 thousand crore at 1996-97 prices. Private consumption expenditure grows at the rate of 6.2 per cent per annum, which implies a per capita consumption growth of 4.5 per cent per annum. At this rate, the per capita consumption level in the country would double in 15 years. More importantly, this scenario sets the stage for even faster growth in the post-Plan period, which could be as high as 7.7 per cent, if not higher, by creating the appropriate level of pipe-line investment in the infrastructure sectors.

6

Public Sector Plan : Resources and Allocations

INTRODUCTION

The Eighth Five Year Plan was formulated with an approved public sector outlay of Rs. 434,100 crore. The anticipated Eighth Plan expenditure is of the order of Rs. 391,000 crore. The lower plan expenditure was the result of the shortfall of about 10 per cent in the resource mobilisation undertaken during the plan period. The projected public sector plan outlay for the Ninth Plan, consistent with the accelerated GDP growth of 6.5 percent per annum envisaged, is placed at Rs.859200 crore at 1996-97 prices. This represents a step up of about 48 per cent and 33 per cent in real terms over the anticipated Eighth Plan expenditure and the approved Eighth Plan Outlay respectively.

This Chapter includes a review of performance in respect of resource mobilisation during the Eighth Plan in order to assess the prospects of resource mobilisation as per the scheme of financing envisaged for the Ninth Plan. The projected scheme of financing is discussed separately for the Centre (including U.Ts without Legislature) and States (including U.Ts with Legislature) followed by a discussion of measures considered necessary to achieve the estimated resources for financing the Ninth Plan.

EIGHTH PLAN PERFORMANCE

The projected financing pattern of the Eighth Plan Outlay of Rs.434,100 crore at 1991-92 prices consisted of budgetary support

(43.4 per cent), Internal and Extra Budgetary Resources (IEBR) of the Central Public Sector Enterprises (CPSEs: 33.2 per cent), and own resources of States (23.4 per cent). This pattern of financing envisaged positive Balance from Current Revenues (BCR) for both the Centre and the States as well as positive contribution by State Level Public Sector Enterprises (SLPSEs). Keeping in view the expected positive BCR and the need to adhere to a non-inflationary scheme of plan financing, the share of borrowings as well as deficit financing in the aggregate resources of the Centre was kept at low levels. However, actual performance of both the Centre and the States revealed significant deviations from projected financing pattern. The areas of shortfalls and deviations are discussed below:

BCR of the Centre and the States

The Eighth Plan Chapter on Financial Resources did not refer to fiscal deficit specifically. However, the need for reducing the reliance on both borrowings and deficit financing was emphasised. Thus the estimates of financial resources for the Eighth Plan included positive BCR both at the Centre and the States. However, contrary to projections, the BCR of both the Centre and the States turned out to be negative during each of the five years of the Eighth Plan. Thus, the combined BCR of Centre and States turned out to be (-) Rs.39563 crore as against the positive BCR of Rs.35005 crore projected for the Eighth Plan. Annexure I to this Chapter provides the details.

The deterioration in the BCR of the Centre was mainly due to (i) decline in indirect tax revenue, and (ii) increased interest burden. The import duty rates were reduced considerably in order to enable Indian industry to reduce cost of production and face international competition. Thus, the customs revenue registered considerable decline from 3.4 per cent of GDP in 1992-93 to 2.7 per cent in 1993-94. The collection rate, i.e the ratio of realised import revenue to the value of imports, is estimated to have declined from 44 per cent in 1992-93 to 37 per cent in 1993-94 and further down to around 30 per cent by 1996-97. However, import growth coupled with depreciation of the rupee reversed this declining trend in revenue from customs which improved from the level of 2.7 per cent in 1993-94 to 3.5 per cent of GDP in 1996-

97, which was only slightly higher than that prevailing in 1992-93. As regards Union excise revenue,the Eighth Plan period witnessed further switch from specific duties to advalorem duties aimed at ensuring buoyancy as measured by the ratio of change in tax revenue to change in GDP at current prices. The scheme of MODVAT was extended with a view to reducing industrial costs and prices by relieving tax on inputs. There was also reduction of rates in the items of mass consumption, white goods, automobiles and synthetic fibres. Exemptions from excise revenue to small scale sector continued. Such exemptions provide loopholes. The net effect of these measures was, a significant drop in excise revenue from 4.4 per cent of GDP in 1992-93 to 3.7 per cent in 1996-97. The main reason for this decline was the growing proportion of MODVAT credit which nearly doubled from about 22 per cent of gross excise revenue in the beginning of the Eighth Plan period to about 42 per cent by the end of the Plan period. Further increase in the ratio of MODVAT credit to gross excise revenue is not expected to be significant as the proportion of revenue from MODVAT to total Union Excise revenue has reached 85 per cent by the end of the Eighth Plan period. Indeed, the proportion of MODVAT credit can be brought down by plugging the loopholes.

As regards direct taxes, the focus of reforms was on rate reduction and simplification to facilitate better compliance. The reform measures resulted in some improvement. The ratio of direct tax revenue to GDP rose from 2.6 per cent in 1992-93 to 3.1 per cent in 1996-97. However, this improvement in direct tax revenue was not sufficient to compensate for the decline in indirect tax revenue mentioned above.

Many items of Non Plan Revenue Expenditure (NPRE) showed declining trend in terms of their proportion to the total NPRE, during the Eighth Plan period. The percentage of subsidies to NPRE came down from 18.2 per cent to 13.2 per cent. However, the proportion of interest payments increased from 39.6 per cent to 46.2 per cent, mainly due to the rise in Government borrowing at market related interest rates.

Balance from current revenues of the States had also declined sharply from Rs. 12985 crore to (-) Rs. 2009 crore in the Eighth Plan. This was mainly on account of substantial shortfall in States'

Own Tax revenues. The realisation in revenue collection was around 85 per cent of the projected level. In absolute terms, the revenue collection was less by Rs. 30,000 crore compared to the estimates of Eighth Plan.

Further, the interest burden also increased from Rs. 10944 crore in 1991-92 to Rs. 26298 crore in 1996-97 (BE) i.e. at an annual compound rate of growth of 19.16 per cent, even though the amount of loans received by the States was only of the order of 89 per cent of the projected level in the Eighth Plan. Rapid growth of interest liability was due to higher interest rates. The interest rate on loans against the small savings varied between 14.5 per cent and 15 per cent during the Eighth Plan period as against 12 per cent to 13 per cent in the Seventh Plan period. Similarly, interest rates on plan and non-plan loans rose to 12 and 13 per cent as against 8 to 9 per cent in the Seventh Plan period. Interest on SLR based market borrowings rose by 2 percentage points in the Eighth Plan period over the Seventh Plan period. This increased burden on account of higher rates of interest exacerbated the deterioration of BCR.

The Eighth Plan – Review of Plan Expenditure and Resource Mobilisation

- The plan expenditure incurred during the Eighth Plan is estimated to be 90 per cent of the total public sector outlay of Rs.4,34,100 crore projected for the five years – 1992-97 at 1991-92 prices.
- Contrary to projections, the Balance from Current Revenues (BCR) of both the Centre and the States turned out to be negative during each of the five years of the Eighth Plan.
- Reduction in import duty rates as well as in excise duty rates for items of mass consumption, white goods, automobiles and fibres coupled with loopholes in the exemptions provided to small scale sector resulted in lower revenue realisation.
- Excise revenue dropped from 4.4. per cent of GDP in 1992-93 to 3.7 per cent in 1996-97.
- Improvement in direct tax revenue from 2.6 per cent to 3.1 per cent of GDP during the Plan period consequent on

direct tax reform, howev er, only partly compensated the decline in indirect tax revenue.

- Non-plan revenue expenditure including subsidies showed declining trend.
- Market related interest rates on Government borrowings, increased interest liability.
- There was substantial shortfall in States' Own Tax Revenue also.
- The interest burden of the States increased at an annual compound rate of growth of 19.16 per cent.
- The resource mobilisation by the Central Public Sector Enterprises (CPSEs) was about 90 per cent of the Plan projections in real terms.
- Failure to undertake price revisions commensurate with the increase in input costs resulted in poor financial performance of the Undertakings.
- Adverse money/capital market conditions affected borrowings of the Enterprises.
- The financial performance of the State Level Public Sector Enterprises (SLPSEs), especially, the State Electricity Boards (SEBs) has been more discouraging. Only three of the SEBs could earn positive rate of return on investment against the statutory obligation for a three per cent rate of return.
- The Centre's dependence on borrowings for financing the Plan increased by 40 per cent over the projections. However, in the case of States, borrowings fell short by 12 per cent.
- The flow of external grant registered only a modest increase whereas there was steep errosion in the net external borrowings mainly on account of increase in the proportion of repayments out of gross borrowings.
- The deficit financing increased by about 65 per cent vis-à-vis the projection of Rs.20,000 crore for the Plan period.

RESOURCES OF PUBLIC SECTOR ENTERPRISES

Internal and Extra Budgetary Resources (IEBR) of CPSEs were projected at Rs. 1,44,140 crore at 1991-92 prices (Annexure 1).

The Actuals for 1992-96 and the Revised Estimates for 1996-97 in respect of generation/mobilisation of IEBR by CPSEs work out to Rs.1,30,324 crore, which is less by Rs.13,816 crore or about 9.6 per cent vis-à-vis the projected estimates. The actual mobilisation of resources by the CPSEs during the first four years of the Eighth Plan has been less compared to the corresponding revised estimates. On this analogy, the actual mobilisation in the fifth year viz., 1996-97 may also be taken to be less than the revised estimates. Keeping this in view, the shortfall in IEBR during the five year period may be placed around 10 per cent of the projected amount.

The CPSEs could not mobilise funds from the market through bonds to the expected level. The adverse money/ capital market conditions and some inherent weaknesses of PSUs were responsible for this shortfall. The failure to undertake necessary price revision to meet the rising input costs resulted in poor financial performance of most of the undertakings. Hence, they could not compete successfully with the private corporate sector for funds in the market. Even the CPSEs which were allowed access to tax-free bonds found it difficult to raise funds from this measure as per target in some of the years because of high market rates of interest. Besides, the allocation of tax-free bonds was restricted even for the CPSEs which were otherwise capable of raising funds through bonds. This was in line with the Government decision to phase out the instrument of tax free bonds. During 1992-93 the amount raised through bonds constituted only around 18 per cent of the target. Though this proportion rose to about 81 per cent in 1993-94, it declined to 41 per cent and 27 per cent in 1994-95 and 1995-96 respectively. The corresponding proportion is estimated to have risen only marginally over the previous year to reach 33 per cent in 1996-97.

In the case of States, the projected internal resources of State Level Public Sector Enterprises (SLPEs) amounted to Rs. 4000 crore. However, actual generation of resources by SLPEs turned out to be negative, i.e (-)Rs. 2723 crore which implied a deterioration of Rs. 6723 crore compared to the projection. This was the result of their poor financial performance, especially of State Electricity Boards (SEBs). The rate of return on investment in State Power Sector continued to be negative throughout the Eighth Plan as against the statutory obligation to achieve a rate of return of 3 per cent. Though there is scope for efficiency gain, the root cause of poor financial performance lies in low tariff which needs to be revised upwards to enable the SEBs to reduce commercial losses.

Borrowings

The deterioration in BCR necessitated greater dependence on borrowings for plan financing. As per the Eighth Plan projections, Centre's borrowings, including Miscellaneous Capital Receipts (MCR), was expected to be Rs. 23551 crore per year whereas the amount borrowed during the Eighth Plan averaged Rs. 33030 crore per annum, which meant an increase of about 40 per cent in the dependence on borrowings vis-a-vis projections. In the case of States, the borrowings including MCR, fell short of projections by around 12 per cent. Taking both Centre and States together, borrowings for financing the Eighth Plan exceeded the projection by about 19 per cent.

Net Inflow of Resources from Abroad

Net inflow of budgetary resources from abroad in the form of external loans and grants amounted to Rs. 19,234 crore as against a projected amount of Rs. 28,700 crore. The proportion of repayments which constituted about 45 per cent of the gross external borrowings in the initial year, increased to about 73 per cent by the terminal year of the Eighth Plan. In nominal terms, the repayments increased from Rs. 4306 crore to Rs. 6969 crore whereas the net external borrowings declined from Rs. 5319 crore to Rs. 2589 crore during the Plan period. Rupee depreciation during this period also contributed to the steep decline. The flow of grants registered only a modest increase from Rs. 919 crore to Rs. 1199 crore during this period. Hence the erosion experienced in net external borrowings could not be made good by the small increase in the external grants.

Deficit Financing

As in the case of borrowings, reliance on deficit financing also exceeded the projections due to the resource gap arising mainly from negative BCR. Besides, minimum budgetary support had to be provided through non plan loans to certain public enterprises to meet cash losses and working expenses. As a result, the recourse to deficit financing in the Eighth Plan was to the extent of Rs. 33,037 crore as against a projection of Rs. 20,000 crore. This meant an increase of about 65 per cent vis-a-vis the projection.

RESOURCE ASSESSMENT FOR THE NINTH PLAN-THE PROFILE

The overall resources for Plan consist of (i) Budgetary Resources of the Centre, including Central Assistance for States and U.Ts., (ii) Internal and Extra Budgetary Resources (IEBR) of the Central Public Sector Enterprises (CPSEs), and (iii) Own Resources of States and U.Ts with Legislature. The size of projected budgetary resources for the Plan depends on the excess of the aggregate projected sum of receipts comprising (a) Balance from Current Revenues, (b) borrowings & other liabilities, (c) other capital receipts (including recovery of loans & advances and disinvestment), and (d) external grants, over nonplan capital expenditure. The IEBR of CPSEs consists of internal and extra budgetary resources. The retained profits and depreciation provision constitute internal resources (IR). The extra budgetary resources include borrowing both at home and abroad, and other resource items like intercorporate transfers, public deposits, external aid received directly from the donors, equity capital raised from the market etc. A part of the aggregate resources of the Centre flows as Central Assistance to the States and the U.Ts. As regards States' own resources, the magnitude depends on BCR, contribution of State Level Public Enterprises, especially SEBs and SRTCs, and borrowings. The Central Assistance mentioned above supplements the States' own resources.

Resources of the Centre

Resources of the Centre consist of both budgetary resources including external assistance routed through the budget and the Internal & Extra Budgetary Resources (IEBR) of Central Public Sector Enterprises (CPSEs). The quantum of budgetary resources of the Centre which is available for providing overall budgetary support to the plan is divided into two parts viz. budgetary support for Central Plan (including U.Ts without Legislature) and Central Assistance for States' Plans (including U.Ts with Legislature). A part of the budgetary resources allocated as budgetary support for the Central Plan is used for providing necessary support to CPSEs. Such budget support to the enterprises has declined significantly from around 50 per cent in mid eighties to below 15 per cent during the Eighth Plan. This necessitated greater dependence of CPSEs on IEBR for financing their plan outlay. As per revised estimates, the proportion of IEBR in the plan outlay of CPSEs ranged from 85 to 90 per cent during the Eighth Plan period. CPSEs in sectors like Petroleum and Telecom depend entirely on IEBR for financing their plan outlay. The provision of budgetary support is limited to (a) CPSEs in the infrastructure sector and (b) loss making CPSEs most of which belong to industry groups like pharmaceuticals, fertilisers, textiles and heavy engineering.

Table 3.1 presents the relevant details on the availability of Budgetary Resources for financing the Ninth Plan.

Table-3.1 : Budgetary Resources for the Ninth Plan (1997-2002)

(Rs.in crore at 1996-97 prices)

Resource Item	Amount
1. BCR	(-)2778
2. Loan recovery	43327
3. Borrowings and other liabilities	391592
4. External Grants	10062
5. Disinvestment	23895
6. Aggregate Budgetary Resources of Centre (1 to 5)	466098
7. Non-Plan Capital Expenditure	92098
8. Centre's Aggregate Budgetary Resources for Plan (6-7)	374000

In the month of April, 1998 the Prime Minister indicated Special Action Plans (SAPs) with the objective of (a) doubling the food production in ten years; (b) expansion and improvement of social infrastructure; (c) expansion and improvement of physical infrastructure and financial services; (d) unveiling a National Water Policy and (e) making India a global information technology power. The Planning Commission was requested to evolve appropriate action plans for achieving the said goals in consultation with the Administrative Ministries concerned and the Finance Ministry.

Accordingly, the Commission formulated Special Action Plans in these areas. These were dovetailed with the normal sectoral plans. The additional financial resources specifically required in respect of the SAPs during the Ninth Plan period were estimated to be of the order of Rs.21,950 crore. The aggregate budgetary support of the Centre to the Plan including the provisions to SAPs is placed at Rs. 374,000 crore.

The magnitude of Centre's aggregate budgetary resources for Plan amounting to Rs. 374,000 crore, as indicated above, is in conformity with the terminal year fiscal deficit of 4.1 per cent of GDP. The projection of BCR shown in table 3.1 given above, is based on tax revenue projections which range from 10.15 per cent of GDP in 1997-98 to 11.5 per cent in 2001-02. The non-tax revenue projected for the Ninth Plan remains at 2.8 per cent of GDP during the plan period. As regards non-plan revenue expenditure (NPRE), the projections take account of the interest liability based on the composition of short-term, medium and long-term loans as well as the monetisation of part of the fiscal deficit. On the average, the rate of interest on borrowings and other liabilities included in the fiscal deficit is estimated at 9.5 per cent. The decline in fiscal deficit from 6.1 per cent of GDP in 1997-98 to 4.1 per cent in 2001-02, has also been considered in projecting the interest liability. The NPRE projected for the Ninth Plan, therefore, shows a decline from 11.05 per cent of GDP in 1997-98 to 9.59 per cent in 2001-02. As regards the capital receipts, loan recoveries were projected to register a negligible growth of about 0.5 per cent at current prices during each year of the Ninth Plan. As regards borrowings and other liabilities, the projections correspond to the targeted level of fiscal deficit in terms of percentage to GDP. Following the Government decision to avoid automatic monetisation of budget deficit through the issue of ad-hoc treasury bills, explicit deficit financing has been kept at zero level.

The Central Government had taken certain measures to meet the impact of the Fifth Pay Commission's recommendations in 1997-98.The annual average burden on this account is expected to be less onerous in the later years of the Plan. The liability in this regard during the remainder of the Plan period has been accordingly included in the projections of NPRE.

The aggregate budgetary support of Rs. 374,000 crore for the Ninth Plan is largely predicated upon a strict fiscal discipline calling for containment of non-plan revenue expenditure and raising the tax-GDP ratio to 11.5 per cent by the end of the Plan. While the former is implicit in the fiscal deficit profile indicated in paragraph 3.19, the latter is mainly a function of tax buoyancy, which consists of the increase in tax revenue arising from both built-in flexibility of and discretionary changes in the tax system. There is considerable scope for raising tax revenue through strict enforcement and expansion of the coverage of Tax Deducted at Source (TDS), as was evident during the post reform period when revenue from corporate income tax as well as personal income tax increased from one per cent of GDP each in 1990-91 to 1.5 per cent and 1.4 per cent of GDP respectively in 1995-96. It is important to recall that the tax-GDP ratio was as high as 11.3 per cent in 1989-90. It declined to 10.8 per cent in 1990-91. Between 1990-91 and 1994-95 it further declined by one percentage point to 9.8 per cent. This decline in the tax-GDP ratio in the first four years of the economic reforms stand in sharp contrast to the increase of 2 to 4 percentage points experienced by some of the developing countries in the wake of their economic reforms. In 1996-97 there has been some improvement, the ratio reaching 10.5 per cent. Considering that Centre's tax-GDP ratio had reached 11.3 per cent in 1989-90 and taking into account the expected impact of discretionary measures referred to above on tax buoyancy, the average tax-GDP ratio projected at 10.67 per cent for the Ninth Plan period is well below the level achieved in 1989-90.

There is considerable potential for mobilising additional resources from direct and indirect taxes through widening of the base, reducing exemptions to the bare minimum and improving the tax administration including stricter enforcement. Measures necessary to exploit the potential are detailed in paras 3.45 to 3.48. In view of this, the projections of tax-GDP ratio mentioned above should be taken as minimal levels that must be achieved rather than as limits of maximal effort.

IEBR of CPSEs

The significant reduction in budgetary support to CPSEs from around 50 per cent of their plan outlay in 1985-86 to about

10 per cent in 1996-97 has made it necessary for them to generate/ mobilise more IEBR for financing their plan outlay. Nearly 90 per cent of the plan outlay of CPSEs is financed through the IEBR generated/ mobilised by them. The IEBR projected for the Ninth Plan amounts to Rs.290,094 crore at 1996-97 prices; an increase of about 25.6 per cent in real terms over the Eighth Plan estimates. This has been estimated on the basis of projected savings of CPSEs and their anticipated borrowings. The projection of IEBR is consistent with the overall public sector outlay of Rs. 859,200 crore envisaged for the Ninth Plan.

Financial Resources for the Ninth Plan – A Profile

- The Ninth Plan (1997-2002) envisages GDP growth rate of 6.5 per cent per annum. The Public Sector Plan Outlay is placed at Rs.8,59,200 crore at 1996-97 prices, representing a step up of 48 per cent and 33 per cent in real terms over the anticipated Plan expenditure and the approved Plan outlay respectively of the Eighth Plan. The share of the Centre's Plan and that of the States, including U.Ts, would be 57 per cent and 43 per cent respectively.
- The Centre's Plan Outlay of Rs.4,89,361 crore is to be financed to the extent of 44 per cent and 56 per cent by way of Gross Budget Support (GBS) and Internal and Extra Budgetary Resoaurces (IEBR) of the CPSEs respectively.
- The Balance from Current Revenues (BCR) is anticipated to be negative as against positive BCR projected for the Eighth Plan.
- The tax revenues are estimated to range from 10.15 per cent to 11.5 per cent of GDP giving an average Tax-GDP ratio of 10.67 per cent during the Plan period.
- Non-tax Revenue will continue to remain at 2.8 per cent of GDP.
- The Non-Plan Revenue Expenditure (NPRE) is projected to decline from 10.05 per cent of GDP in 1997-98 to 9.5 per cent in 2001-02.
- Borrowings are likely to contribute 60 per cent of the financial resources for the Centre's Plan.

- The CPSEs are expected to finance 90 per cent of their Plan Outlay through IEBR.
- The fiscal deficit is expected to decline from 6.1 per cent to 4.1 per cent of GDP during the Plan period.
- Borrowing by States will constitute 47 per cent of their Ninth Plan Outlay as in the projections for the Eighth Plan.
- The Internal Resources of the State Level Enterprises are also expected to be higher at 4 per cent of the Plan Outlay as against 2 per cent projected for the Eighth Plan.
- The States' BCR (including Additional Resource Mobilisation – ARM commitment) will improve to become positive during the Ninth Plan.

The financial sector reforms have enabled the corporate sector including CPSEs to raise resources from the market both at home and abroad. Howsever, the public sector enterprises have to compete with their private sector counterparts for raising resources. This necessitates more autonomy for CPSEs to enable them to improve their financial performance. CPSEs are required to maximise generation of IR during the Ninth Plan period. Efficiency gains through higher productivity alone cannot achieve this. It is necessary to facilitate periodic price revision to meet the rising input costs. With improvement in operational efficiency and adoption of suitable pricing policies, it is expected that the CPSEs would be able to raise the projected level of IEBR during the Ninth Plan.

Financing Pattern of Central Plan Outlay

The financing pattern of the Central Plan Outlay inclusive of IEBR of CPSEs is given in Table 3.2.

It will be seen that the financing pattern for the Ninth Plan is markedly different from the earlier pattern in as much as deficit financing has been kept at zero level and balance from current revenues has turned negative. This has necessitated considerably higher percentage of borrowings and other liabilities.

The Gross Budgetary Support to the Central Plan for the first two years of the Ninth Plan (1997-98 – Revised Estimates and Budget Estimates for 1998-99) accounted for only about 15.4 per cent and 17.1 per cent respectively of the total such support for Ninth Plan. Consequently, the budgetary support for the remaining three years will have to be considerably enhanced so as to achieve the projected Central Plan Outlay for the Ninth Plan. This calls for stepping up of the resource mobilisation efforts on the lines indicated in paras 3.45 to 3.48. As regards IEBR also, the Revised Estimates for 1997-98 and Budget Estimates for 1998-99 work out to about 15.6 per cent and 19.2 per cent respectively of the total for the Plan period. Therefore, in order to finance the projected plan outlays, the CPSEs will be required to raise higher levels of IEBR in the remaining years. Annexure II shows budgetary support, IEBR and plan outlay of Central Ministries/ Departments for the Ninth Plan.

Table 3.2 Financing Pattern of Central Plan Outlay (including U.Ts without Legislature)

Resource	Amount (Rs.in crore at 1996-97 prices)	percentage share in aggregate resources
1. BCR	(-)2778	(-)0.4
2. IEBR of CPSEs	290094	43.7
3. Borrowings and other liabilities	391592	59.0
4. Misc. Capital Receipts(net)	(-) 24876	(-)3.7
5. External Grants	10062	1.5
6. Deficit Financing	0	0.0
7. Aggregate Resources(1 to 6)	664094	100.0
8. Assistance for State Plans	(-)171873	(-)25.9
9. Resources for Central Plan *	492221	74.1

*includes UTs without Legislature

Notes: 1.Borrowings and other liabilities include External Borrowing of Rs.49956 crore.

2.Misc.Capital Receipts(net) consist of the difference between non-debt Capital Receipts and Non-Plan Capital Expenditure.

3.Resources for Central Plan include an amount of Rs.2860 crore as Central Assistance for UTs without Legislature.

The financing pattern of the Centre and States and the gross budgetary support to the Plan are based on the recommendations of the Tenth Finance Commission. In the computation of States Own Resources and the BCR of the Centre, it has been assumed that 29 per cent of Centre's gross tax revenue will be devolved to the States after passing the requisite Constitution Amendment Bill. However, the Tenth Finance Commission award will be valid only until 1999-2000. The resource position may be significantly altered depending upon the recommendations of the Eleventh Finance Commission.

RESOURCES OF STATES

The resources of the States for their Plans consist of States' Own Resources (SOR) and Central Assistance (CA). The SOR includes balance from current revenues (BCR) of the States, internal resources of their enterprises, mainly of the State Electricity Boards and State Road Transport Corporations, devolution of grants through Tenth Finance Commission for capital works, borrowings by State Governments and their enterprises and miscellaneous capital receipts (MCR). The Central assistance includes domestic budgetary assistance as well as assistance for externally aided projects.

As mentioned earlier (paras 3.4 to 3.8), resource mobilisation during the Eighth Plan fell short of projection. Table 3.3 shows that the total resource mobilisation for financing the Eighth Plan of States constituted only about 81 per cent of the projected amount.

The main shortfall was in respect of BCR and contribution of the public enterprises. As against the total amount estimated at about Rs. 17000 crore from these two sources, the realisation would be around (-) Rs. 4730 crore leading to a deterioration of about Rs. 21700 crore. There was also substantial shortfall in the assistance for Externally Aided Projects of the order of Rs. 5400 crore. The entire SOR of States at the aggregate level accrued from borrowed funds, some part of which was utilised to meet non-plan revenue expenditure and negative contribution of their enterprises. The amount borrowed by the States for financing their Eighth Plan did not exceed the projection. However, these borrowings were at higher rates of interest. Further these were largely utilised to finance social sector outlays where there are no direct and immediate financial returns. Even where such funds are invested in productive sectors like power, irrigation and transport, the financial returns are extremely poor. All this has resulted in compounding the debt servicing problem of the States, the effect of which will be more evident in the coming years. Were the projected levels of borrowings to be reached, the plan outlay would have been higher. However, this would have exacerbated the debt burden.

The shortfall in resource mobilisation indicated above reinforces the conclusion that States will have to take concrete steps to improve their balances from current revenues and generation of internal resources of their enterprises. Improvement in BCR also presupposes compression of non-plan revenue expenditure.

RESOURCE ASSESSMENT FOR THE NINTH PLAN

The estimates of financial resources for the Ninth Plan were firmed up during the discussion between Deputy Chairman and the Chief Minister. The final position that emerged is presented in Table 3.4.

The BCR estimate of the Eighth Plan projections was too optimistic accounting for 7.22 per cent of total resources. In reality, however, the BCR turned out to be significantly lower reducing its share to (-) 1.36 per cent. Keeping this in view and the likely additional liability on account of pay revisions, the Ninth Plan projection of BCR including the budgetary ARM has been kept at a modest level accounting for 0.39 per cent of the total resources. The Eighth Plan projection estimated the share of contributions by State enterprises at 2.22 per cent of the total. In fact, it turned out to be (-) 1.86 per cent. For the Ninth Plan, the share of contribution of State Enterprises including ARM is estimated at 4.25 per cent of the total. This implies a thorough re-organisation and revamping of the structure and functioning of the SEBs and SRTCs. In the projection for Eighth Plan the share of borrowing was a little less than 47 per cent. While the actual amount of borrowing fell short of projections its share in the total resources as realised was much higher at, 51.37 per cent. This was mainly because the performance in respect of BCR and contributions of PSEs was far lower than the projections. The share of borrowing in the total resources in the Ninth Plan is kept slightly below the level experienced in the VIII Plan. The share of Central Assistance in the States' Eighth Plan projection was about 44 per cent. The share of Central Assistance for States in the Ninth Plan is kept slightly higher than the projected level in the Eighth Plan.

As has been observed earlier, the scheme of financing of Ninth Plan of the States visualises substantial efforts for mobilisation of additional resources through budgetary measures as well as through measures to be taken by State enterprises. The States have to take a number of steps to improve their tax GDP ratio. These include tax reforms, focussing on rationalisation of tax structure, specially the Sales Tax. The introduction of prohibition in some States had a softening impact on raising resources. This needs to be made up through compensatory measures in the other elements of tax structure and improvement in tax administration. Equally important is the need for empowering the States to tax services sector, to impose consignment tax and also to revise royalty rates for major minerals as per the recommendations of the Sarkaria Commission.

The SEBs are statutorily required to realise 3 per cent return on net fixed assets. In the Eighth Plan the average realisation of all SEBs was of the order of (-) 9.5 per cent. Out of 19 Boards, 15 had negative returns and 4 had positive returns of which only 2 had rate of return of above 3 per cent. The main reason for their extremely poor performance has been their inability to rationalise tariff structure to recover at least the cost of supply. The operational inefficiency and transmission and distribution losses are the other reasons. It is expected that concerted efforts will be made to restructure the SEBs, revise the tariff and improve their operational efficiency during the Ninth Plan period.

The functioning of the SRTCs need to be substantially improved. The fare structure should be revised periodically to yield reasonable returns. These Corporations have a number of social obligations to meet. It will be necessary to ensure that the Corporations continue to have remunerative routes so that they can cross subsidise the social obligations.

At the time of discussions with State Chief Ministers to finalise the resources for the State Plans, it had been estimated that the additional liability for pay revisions would be of the order of Rs.60,000 crore.

However, the latest estimates indicated by the States place this burden at around Rs.11,0000 crore. The extra burden is unlikely to be met entirely by ARM. It is likely that the BCR of the States will not improve to the extend desirable and continuing reliance will have to be placed on market borrowings. It is, therefore, assumed that the share of market borrowings in the scheme of financing of the States' plans will continue at the level experienced during the VIIIth Plan.

Sufficient attention has not been paid to realisation of non-tax revenues by the States. The areas from where the non-tax revenues can be augmented are irrigation charges, royalties on minor minerals and revision of user charges on services rendered by the Government. There is an urgent need for more frequent revisions of water rates and effective water rate administration atleast to meet the full O&M expenditure of the irrigation department.

The Central Assistance to the States' Plan is projected at Rs. 169152 crore comprising of Rs. 129152 crore of domestic budgetary support and Rs. 40000 crore of assistance for externally aided projects (EAPs).

The Central Assistance to the States in 1997-98 amounted to Rs. 24192 crore at 1996-97 prices. This works out to only about 15 per cent of the Ninth Plan. It follows that the Central Assistance has to be substantially stepped up in the remaining years of the Ninth Plan.

Some of the measures to improve the State finances like alternative scheme of devolution recommended by the Tenth Finance Commission, revision of royalty rates, introduction of consignment tax, delegation of powers to the States to tax service sector, setting up of Tariff Regulatory Commissions for power sector etc. require the support from the Centre. The measures to be taken by the Centre to supplement the resources of the States are mentioned in detail in the section on measures to be adopted by the Centre to improve the resources of the States.

Union Territories

The outlay for Union Territories during Eighth Five Year Plan was projected at Rs.6250 crore. The anticipated expenditure amounts to Rs.6530 crore indicating a performance level of 104.5 per cent.

The resources of the UTs without legislature are equal to the Central Assistance. In the case of other two UTs with legislatures, there is the additional element of their own resources. Table 3.5 shows the resource position of UTs for the Ninth Plan.

Overall Financing Pattern

The aggregate public sector resources for the Ninth Plan work out to Rs.859200 crore. This comprises of Centre's resources of Rs.664094 crore including Central Assistance to States and UTs (vide item No.8 of Table 3.2), States' own resources of Rs.180986 crore (vide item No.4 of Table 3.4) and U.Ts own resources of Rs.14120 crore (vide Table 3.5). The Table 3.6 presents the scheme of financing the plan outlays of Centre, the States and the U.Ts.

7

Employment Perspective

LABOUR FORCE GROWTH AND EMPLOYMENT REQUIREMENTS

One of the most daunting challenges facing the Ninth Plan is to provide employment not only for the additions to the labour force during the Plan period, but also to reduce the back-log of unemployment accumulated from the past. This imperative needs to be seenin the context of both an acceleration in the growth rate of the labour force and a secular downward pressure on the employment intensity of the growth processes.

Despite an expected reduction in the growth rate of population to 1.58 per cent per annum by the end of the Ninth Plan, the labour force growth reaches a peak level of 2.51 per cent per annum during the Ninth Plan period; the highest it has ever been and is ever likely to attain. Population is projected to increase by 78 million during the Ninth Plan period (1997-02) to a level of 1029 million by the end of the Plan.

However, the age structure of the population changes in response to fertility behaviour experienced in the past, and projected for the future, so that the proportion of the population in the most active working age-group of 15 to 59 years goes up significantly during the Plan period. This is a reflection of the high rates of population growth observed during the late 1970s and early 1980s. Moderation in growth of population, which has been brought about by a decline in birth rate in recent years, starts getting reflected in decline in labour force growth only in the post Ninth Plan period.

The growth rate of the labour force is determined partly by the age structure of the population and the age and sex specific labour force participation rates that can be expected to obtain.

The age profile of the population for the Ninth Plan period is presented in Table 1.

As can be seen, the Ninth Plan period is likely to witness a sharp decline in the share of the population in the age group of 0 to 14 years, and a corresponding increase in the age group of 15 to 59 years, with the 60+ age group more or less retaining its share.

Table 1 : Age structure of population: 1997-2002

(Per cent)

Age-group	1997	2002
0-14	37.23	33.59
15-59	56.07	59.41
60+	6.70	7.00

The labour force participation (LFPR) by age and sex are given in Tables 2 & 3. As may be seen, participation in labour force varies across age-groups, rising sharply after 20 years age in case of males and 34 years in case of females.

Furthermore, for any particular age-group, it tends to vary over time, the direction of change depending upon the benefit perceived by the individual from devoting time to an economically productive activity in contrast to pursuit of education or leisure. A crucial factor in the choice between joining and remaining in the labour force or otherwise is the availability of social security, either through the family or through other institutional structures.

The most recent year for which labour force participation rates (LFPR) are available by age-groups is 1993-94. Variations in LFPRs during the past 16 year period are not uniform across age-groups; there is a substantial decline in LFPR in age group 15-29, practically no change in 30-44 years, and a slight increase in the higher age groups.

Table 2 : Trends in Labour Force Participation Rates

(Per Thousand of Population)

Age Group	Period	Male		Female	
		Rural	Urban	Rural	Urban
15-29	1977-78	879	746	515	257
	1987-88	824	710	478	211
	1993-94	804	684	455	204
30-44	1977-78	990	990	619	324
	1987-88	988	987	603	301
	1993-94	990	986	600	300
45-59	1977-78	963	940	538	291
	1987-88	964	933	538	275
	1993-94	968	937	543	283
60+	1977-78	667	517	221	130
	1987-88	670	482	220	123
	1993-94	699	443	241	114
All(15+)	1977-78	904	831	517	269
	1987-88	879	810	496	239
	1993-94	877	811	491	238

Note: Constituent shares in labour force in 1993-94 are Rural Male 0.499, Rural Female 0.270, Urban Male 0.182 and Urban Female 0.049.

A continuation of the trend decline in labour force participation is projected in the age group 15-19 years over the perspective period. The decline may be even sharper if the thrust on education and literacy that is proposed for the Ninth Plan is realised to a substantial extent. As far as the base year (1996-97) projection of the LFPRs for the other age groups are concerned, these have been derived on the basis of the information pertaining to the 1987-88 and 1993-94 NSS data and the 1991 Census. It turns out that there is practically no difference between these figures and the 1993-94 NSS data. Assuming no significant change in participation rates occurring in other age groups in relation to the base year, the LFPRs by age and sex are given in Table 4.3. However, it is recognised that with improved health and longevity, the

LFPRs in the older age groups, particularly 50+ years, may be higher than assumed. On the other hand, with improvements in the average economic conditions, the imperative to remain employed for survival reasons is mitigated. Thus there are contradictory forces operating in this regard, and it is implicitly being assumed that the net effect is not significant.

Table 3: Participation in Labour Force by Age Group and by Sex: 1997-2012

(per thousand of populaltion)

Age	Males				Females			
	1997	2002	2007	2012	1997	2002	2007	2012
15-19	517	482	447	412	302	282	261	241
20-24	871				408			
25-29	975				454			
30-34	988				505			
35-39	990				526			
40-44	986		(a)		538		(a)	
45-49	981				524			
50-54	961				476			
55-59	914				411			
60 +	637				205			

Note: (a) No change in labour force participation in age groups above 20 years.

On the basis of the age and sex structure of the population and the LFPRs, the labour force is projected to increase by 52.4 million during the Ninth Plan period on the usual status concept, as shown in Table 4. As may be seen, the absolute number of persons in the 35-60 age group of labour force increases sharply during the Ninth Plan because of the change in age structure, which leads to 3 percent or higher growth occurring in labour force age groups 40 to 59, despite an expected reduction in the LFPR of this age-group. If the motivation to attain higher education levels or the expected level of economic well-being and social security do not materialise, the LFPR's of lower age groups may turn out to be higher and put even further pressure on the labour market.

Table 4 : Labour Force Projections by Age Groups

Age Group	1997 (Million)	2002	Growth (% p.a.)
	15-19	40.31	45.03 2.24
20-24	55.45	62.91	2.55
25-29	56.89	61.47	1.56
30-34	52.64	58.88	2.26
35-39	46.60	52.80	2.53
40-44	39.56	46.04	3.08
45-49	32.90	38.13	2.99
50-54	25.86	30.27	3.20
55-59	18.86	22.45	3.55
60 +	28.15	31.64	2.37
15 +	397.22	449.62	2.51

The labour force growth corresponding to the projected increase in population, change in its age structure and change in labour force participation for the perspective period is given in Table 5.

It will be seen that the requirement of work opportunities increases by 52 million in Ninth Plan, followed by an increase of 58 million in the quinquennial 2002-07, and a mild reduction to 55 million in quinquennial 2007-12. Thus, although the rate of growth of the labour force is highest during the Ninth Plan period, the increase in the absolute number of persons in the labour force will peak during the post Plan quinquennium of 2002-07. Efforts will therefore have to be initiated for creating the conditions for absorption of the larger number of workers in the post-Plan period during the Ninth Plan period itself.

Table 5: Population and Labour Force: 1997-2012

(million-Ist April)

	1997	2002	2007	2012
Population	951.18	1028.93	1112.86	1196.41
Labour Force	397.22	449.62	507.94	562.91

EMPLOYMENT OPPORTUNITIES DURING THE NINTH PLAN AND BEYOND

In measuring employment / unemployment in a country like India, certain specific features of the workforce need to be taken into account. The structure of workforce with dominance of self-employment and primary sector, where work-sharing is common, also tends to depress unemployment rates in general and chronic, long-period unemployment rates in particular. Inadequacy of the measure of unemployment in terms of open unemployment has, therefore, been well recognised in the measurement and analysis of unemployment in India. The National Sample Survey Organisation (NSSO), which provides estimates of the rates of unemployment on the basis of its quinquennial surveys, therefore, uses three different concepts. A person is considered unemployed on Usual Status (US) basis, if he/she was not working, but was either seeking or was available for work for the major part of the reference year. On the basis of a week as the reference period, a person is considered unemployed by Current Weekly Status (CWS), if he/she had not worked even for one hour during the week, but was seeking or was available for work. The third concept of unemployment is the Current Daily Status (CDS), which is in terms of total person days of unemployment, and is the aggregate of all the unemployment days of all persons in the labour force during the reference week. The US unemployment rates is generally regarded as the measure of chronic open unemployment during the reference year; the CWS unemployment rates also measure chronic unemployment, but with the reduced reference period of a week. The CDS is considered to be a comprehensive measure of unemployment, including both chronic and invisible unemployment.

Of the three concepts used to study disposition of time in order to arrive at estimates of labour force and employment-usual status, current weekly status and current daily status-the usual status concept covering the principal and subsidiary workers is the closest to the concept used in the Census to enumerate workers. By including even those who are not principal workers but work in a subsidiary capacity, e.g., students, pensioners, etc., the usual status concept comprehensively covers all those engaged in or seeking economic activities. In a situation where social security in the form of a reasonable and assured support as allowance for unemployment is not guaranteed, those who take to work in a subsidiary capacity do so generally out of economic compulsion. Hence both the Principal and the Subsidiary workers need to be considered in estimating both the labour force and employment. Moreover the usual status concept; with one year as the reference period over which disposition of time to activities is seen in classifying an individual as being in or out of labour force, and those in labour force as being employed or seeking work, yields more stable estimates of employment in contrast to the other two concepts which have shorter reference periods of a week or an average day of the week respectively. Taking the usual status concept as the reference, it is possible to study the nature and characteristics of those employed, using the information available from the data compiled on current weekly and current daily status concepts bases. Projections of labour force and employment have therefore been made on the usual status concept, and qualified, where necessary, on the basis of the other two concepts.

Given the existing base of information, two independent and unrelated sets of data have to be used for estimating the relationship between employment and output : (i) estimates of employment and unemployment prepared on the basis of allocation of time by an individual to different activities; and (ii) estimates of output made using national income concepts. Data for the former is collected through quinquennial National Sample Surveys, and for the latter through the system of National Accounts. Although there are some conceptual differences between the two sources of data, they can be used so long as the differences remain consistent over time. The relationship between the two variables, i.e. output and employment, at the sectoral level has varied over time, but has been observed to have remained within a reasonable range if the effect of economic shocks, such as droughts, are accounted for. Thus it is possible to econometrically estimate with some margin of error, the elasticity of employment on a sectoral basis which can then be utilised for projecting employment both at the sectoral and at the aggregate level for the Ninth Plan period. These elasticity estimates are provided, but it needs to be recognised that these estimates can change depending upon the economic structure and composition of growth at the sub-sectoral level.

As has been noted, the growth of labour force accelerates in the Ninth Plan period in comparison to the preceding decade, which requires a commensurate increase in the pace of creation of additional work opportunities in this period. This factor needs to inform every aspect of formulation of the Ninth Plan.

The growth target for Ninth Five Year Plan needs to be 6.5% per annum on an average. The sectoral structure of growth corresponding to the target of 6.5% growth in GDP, as given in Chapter 2, along with the estimated sectoral elasticities, yield an increase in work opportunities during the Ninth Plan period as shown in Table. The projections of labour force and work opportunities created during the Ninth Plan lead to a reduction in the average rate of unemployment during the Ninth Plan period as compared to the Eighth Plan.

Table : Projections of Work opportunities 1997-2002

	Sector	GDP Growth (% p.a.) (1997-02)	Work opportunities (Million) 1997	2002
1.	Agriculture	3.9	238.32	262.48
2.	Mining & Quarrying	7.2	2.87	3.54
3.	Manufacturing	8.2	43.56	48.22
4.	Electricity	9.3	1.54	1.93
5.	Construction	4.9	14.74	17.03
6.	Wholesale & Retail Trade	6.7	34.78	41.67
7.	Transport, Storage & Communication.	7.3	11.96	14.57
8.	Financing, Real Estate, Insurance and Business Services	8.5	4.55	5.68
9.	Community, Social and Personal Services	7.1	38.98	46.41
	All Sector	6.5	391.30	441.52

Table : Labour Force and Employment in Eighth and Ninth Plan Periods

(Million)

	8th Plan(1992-97)	9th Plan (1997-02)
Labour Force	374.2	423.4
Employment	367.2	416.4
Unemployment	7.0	7.0
(rate) (Percent)	(1.87)	(1.66)

Every effort would need to be made in attaining full employment in the post-Plan period. Since labour force growth is expected to be at its peak in the Ninth Plan period, attainment of near full employment by the year 2007 may not be an unreasonable target provided that the conditions are created for further acceleration in the growth rate and the intensity of labour absorption is not substantially reduced both sectorally and in terms of the sectoral structure of growth. It is estimated that full employment by 2007 is contingent upon acceleration in growth of employment in post Ninth Plan period to 2.8 per cent compared to a realised growth of 2.36 per cent (1978-94) and projected at 2.44 per cent in the Ninth Plan. This would require GDP to grow at 7.7 per cent per annum during the post-Plan period-which is well within the realm of feasibility, as shown in Chapter 2.

8

Implementation, Delivery Mechanism and Institutional Development

INTRODUCTION

During the last 50 eventful years, India has acquired rich experience and learnt useful lessons in the field of economic development and social change.

When the First Five Year Plan was launched, the planners were aware that they were super-imposing economic development programmes on a semi-feudal society, which was characterised by age-old social inequalities, long persisting economic disparities and a general attitude of the common masses to expect everything from the Government. Besides, the Indian economy was saddled with outmoded legal institutions, incompatible administrative system and procedures and generally unenterprising behavioural attitude of even those who were functioning as economic agents. The nation decided to launch upon the development plans with the hope that some of these institutional barriers which were operating in the Indian economy and hindering economic development and social change might get themselves transformed and become positive motivating factors. They were also aware of the strength of the deep-rooted institutional deficiencies and proceeded with the assumption that these institutional barriers could be rectified as the development process proceeded along and the deficiencies could be minimised as and when found necessary and feasible in the course of implementing the development plans. The past fortyfive years of development experience reveals that while some minor institutional hurdles have been gradually weakened by the forces of development initiated through successive Five Year Plans, the nation has not fully succeeded in building up institutions to answer the needs of the times.

Many institutional hurdles and semi-feudal socio-economic institutions, which existed at the time of Independence, have remained. Notwithstanding these hurdles, deficiencies and failures, the Indian economy has entered into a higher growth path as compared to not only the pre-Independence period but also the early Five Year Plan periods. Covering a span of thirty years from 1951 to 1981, the Indian economy grew at the rate of 3.5%, which of course, was more than the growth of its population and much more impressive compared to the decadent economic situation in the pre-Independence decades. During the eighties, the Indian economy entered into the higher growth path of nearly 6% per annum and the growth performance has become even better during the recent years. This is not a small achievement considering the vastness of the country and disparate regions. But what is of concern is that these achievements have not been uniform across regions, sectors and groups of people. There lies the inherent weaknesses of the strategies and programmes of Indian Five Year Plans.

This Chapter takes stock of the implementation of the Five Year Plans with a view to identifying the weak spots in the formulation and implementation of the Plan programmes and projects and will attempt to find solutions to the weaknesses in the delivery system. Implementation of the Plan involves designing of the strategies, programmes and schemes, designing of the projects, monitoring of their implementation and ex-post evaluation of the results or benefits flowing from the projects.

IMPLEMENTATION & DELIVERY SYSTEMS-A REVIEW

For the realisation of the intended Plan objectives, the quality of design and effectiveness of implementation of the Plan programmes are as important as the availability of resources. Past experience has shown that many development projects and programmes, having laudable objectives, have failed to deliver the result/ because of the inadequacies in design and implementation. Time and cost overruns have been widespread and substantial in public sector infrastructure and investment projects. It is common knowledge that the benefits intended to be delivered to the people through development programmes in the social sectors have not fully reached the beneficiaries because of the weakness in administrative planning and delivery mechanism. An area of priority of the Ninth Plan is to improve project planning and implementation to ensure that avoidable time and cost overruns are reduced, if not eliminated and the intended benefits actually reach the people.

While designing and executing projects/ programmes, assumptions are made about some technical and non-technical parameters because of imperfect knowledge about the local situation, inadequate data base and time and resource constraints. In designing large irrigation projects, for example, assumptions are made about water availability, land acquisition, irrigation intensity, cropping pattern, land productivity and co-operation of the water users-all of which have a bearing on the viability of projects. More often, such assumptions turn out to be unrealistic or too optimistic and, as a result, the progress of implementation of such projects suffers. This has been corroborated by ex-post evaluations of such projects. While it is possible to bring about an improvement in project design and implementation through extensive groundwork and giving adequate attention to the minute details of every aspect, it is not always feasible because of time and resource constraints and certain unforeseen factors.

A more practical approach to programme/project formulation and implementation is to put the lessons learnt from experience to practice. Ex-post and concurrent evaluation of projects have brought out how the inadequacies in design and implementation affected the completion of a project. In the past, these lessons were not fully systematised and adequately used by the policy makers, planners and project managers. To ensure effective utilisation of available resources, the Ninth Plan strategy for programme implementation and delivery system will be built upon the lessons learnt from experience in the execution of innumerable projects and programmes in the past. To this end, a review of the past experience has been made to identify the factors that generally contributed to delay in completion and sub-optimal project outcomes. As the nature and problems of implementation of public sector infrastructure and investment projects differ from those of poverty alleviation and social sector programmes, the lessons learnt from experience and the steps to be taken for efficiency improvement are discussed under separate headings.

PUBLIC SECTOR DEVELOPMENT PROJECTS

A recent review of 187 Central sector projects, each costing more than Rs.100 crore, by the Department of Programme Implementation (DPI) showed that, as on February, 1997 as many as 118 projects were running behind schedule and the time overrun varied from one month to 200 months.

A study of a few major projects indicated that the cost overrun, for reasons other than inflation and changes in the duty/ exchange rate regimes, ranged from 40% to 75% of the original estimates. Past experience reveals that the factors responsible for time and cost overruns are:

- Poor project formulation due to inadequate field investigation, lack of adequate data, inadequate analysis of environmental and rehabilitation implications, changes in prices and exchange rate regimes, etc.
- Delays in clearance from various regulatory agencies in land acquisition and in procurement of materials. Such delays are primarily due to poor coordination and project planning, as these problems are not explicitly considered or taken into account at the planning stage.
- Changes in design or scope of projects midway through execution.
- Inability of the project management to take prompt decisions on various aspects of these projects even when the objective circumstances warrant such decisions.
- Management problems such as personnel, labour and contractor disputes, mis-match of equipment, etc.
- Inadequate and untimely release of funds.

- Unforeseeable factors such as adverse geo-mining conditions and natural calamities.

Reviews of some projects with substantial time and cost overruns were undertaken to identify activity-wise delays, type of cost overruns and the reasons thereof. In a coal washery project with a time overrun of 10 years, the activity-wise time overruns and the reasons thereof were: 4½ years in land acquisition owing to litigation and law and order problem; 1½ years in land filling work (not envisaged at the planning stage) due to poor planning; and 4 years due to non-fulfillment of contractual obligations. About 94% of the cost overrun of the project is explained by these delays.

An ongoing power project, which was to be completed in 6½ years, has already taken more than 11 years to complete about 90% of the project activities. The expenditure incurred so far is about 210% of the original estimates. Inordinate delay in land acquisition, inability of a public sector unit to supply vital equipment and designs as per contract, slow progress of work by private contractors, changes in scope midway through execution, serious funds crunch on account of non-payment of dues by electricity boards, inadequacies in the tender documents etc. are found to be the major reasons for time and cost overruns.

Time and cost overruns due to changes in design and scope midway through execution are quite frequent in large public sector projects. In the case of large surface irrigation projects, in particular, this problem is so acute that several projects had alarming time and cost overruns. To cite an example, an irrigation project in a State, which was initially designed in 1953 for flood protection and irrigation benefits to 5.7 lakh hectares, underwent drastic changes in design and scope, midway through execution, to include: (a) hydel power generation, (b) an additional branch canal with a command area of 1.6 lakh hectares, (c) another main canal system taking off from the barrage with an aggregate command area of 3.25 lakh hectares, (d) extension of the embankments (flood protection) and (e) cost of watercourses as the project cost. Because of these changes, the project got delayed by more than 20 years and the cost escalation, as per the latest available estimate was 347.25 percent. It is also not clear if the benefit-cost ratio of the entire project will still be favourable, as the planned power generation and irrigation capacities have not been realised.

It is quite common in large projects, particularly in mining and large scale surface irrigation schemes that the project authorities have not been able to get possession of land in time for the construction of various components of the project. Generally, two factors are associated with the problem of land acquisition. First, more often than not, there is a significant difference between the land prices assumed at the time of project planning and the market rates, as a result of which the process of land acquisition leads to time and cost overruns. The second problem relates to the absence of an institutional mechanism for speedy disposal of disputes arising in the context of land acquisition. When such disputes acquire a high profile, as in the case of Tehri and Narmada dams, the projects get delayed indefinitely with adverse consequences on costs and benefits.

The findings of diagnostic evaluation studies/ reviews lend support to the general observations about the factors causing time and cost overruns. However, what is important to note is that factors like problems relating to land acquisition/rehabilitation, obtaining clearances, non-fulfillment of contractual obligations by both public sector units and private contractors, inadequate and untimely release of funds and inadequacies in tender documents contribute more often to the greater part of the time and cost overruns of public sector projects. These problems arise due to inadequacies in planning procedures and implementation.

Project planning has to be more scientific and approval procedures more realistic to ensure that avoidable time and cost overruns are much less frequent. The approval procedure should be linked with early completion of incomplete projects and sustainability of project output. Because of the unrealistic approval procedure, many of the projects are delayed. At the other extreme, less stringent approval procedures encourage a tendency to get too many projects cleared without the requisite financial resources in sight. There is, thus, a need for striking a balance between these extremes. It is important to ensure that rigour in appraisal and planning does not itself become a cause of delay because of repetitive and multi-level examination of technical and economic data. Strict time-tables need to be laid down for completion of the approval processes and preliminary work. Similarly, strict financial procedures should be formulated for eliminating projects, which do not have financial backing.

The implementation of projects will considerably improve if the planning and approval procedures become more scientific and rigorous. However, for effective project management, there is a need to keep track of the progress in implementation and take necessary corrective actions, as the progress may be affected by unforeseen factors. Thus, monitoring and evaluation system must be strengthened and the implementing agencies must be made accountable for non-adherence to the plan of work. There is evidence that effective monitoring of implementation and mid-course corrections by the project management have often led to timely completion of projects without any cost overrun. An IFFCO fertilizer project was completed well within the approved cost estimate and time schedule, even though there had been delays in certain activities ranging from one to seven months during implementation. This was possible because the project management used to review the progress every week and take corrective actions to re-adhere to the original schedule of works. The corrective actions, inter alia, included actions against defaulting contractors, provision of standby generators, import of steel structures and cement to augment domestic supply and the like.

Apart from rigour in planning and project management, certain procedural and institutional reforms wouid be required to reduce avoidable delays and thus cost overruns. Time-bound clearances at different stages and effective inter-agency coordination would cut down time and cost overruns considerably. It is also not uncommon to find that where public sector units are the main providers of engineering design, equipment and material, non-fulfilment of contractual obligations by these agencies has been a major source of delays in project completion. Appropriate rules and procedures must be framed to avoid such delays and alternative strategies to deal with such situations by the project management must be evolved. The project management must be made fully responsible for the inadequacies in tender documents which often lead to delays, avoidable litigations and cost escalations.

Stringent penal clauses to deal with non-compliance must form a part of the tender documents and suitable laws must be in place for early resolution of legal disputes.

Sustainability of the assets created has received much less attention in the planning process. This is the primary reason for the deteriorating conditions of assets and low capacity utilisation. The project design should include the parameters which bring out clearly as to how the project output in the form of goods and services will be sustained during operation, how various assets will be maintained and quality ensured. This question is also linked with the issue of 'pricing' of project output. This too must be squarely addressed, and within some defined parameters the project authorities should have freedom to price their output to recover legitimate costs and generate a reasonable surplus for further investments.

9

Cooperative Federalism and Decentralisation

INTRODUCTION

In a vast country like ours, the spirit of co-operative federalism should guide the relations between the Centre and the States on the one hand, among different States and between the States and the Panchayati Raj Institutions (PRIs) and the Urban Local Bodies (ULBs) on the other. The essence of co-operative federalism is that the Centre and the State Governments should be guided by the broader national concerns of using the available resources for the benefit of the people. Co-operative federalism encourages the Government at different levels to take advantage of a large national market, diverse and rich natural resources and the potential of human capabilities in all parts of the country and from all sections of the society for building a prosperous nation.

Co-operative federalism makes it possible to raise all the available resources by the Government at different levels in a co-ordinated way and channel them for use for the common good of the people. This requires a harmonious relationship and co-operative spirit between the Centre and the States and among the States themselves. While a healthy competition among the States for evolving efficient and socially desirable policies and programmes is welcome, any competition which nullifies each other's advantages in development and erodes the resource base of the States should be avoided. Co-operative federalism is intended to ensure a minimum bundle of basic services and a nationally acceptable level of living for all the people of the country.

CENTRE-STATE RELATIONS

The relations between the Centre and the States in political, economic, financial and administrative spheres have been periodically reviewed. The Administrative Reforms Commission and the Sarkaria Commission were appointed to review the whole gamut of relations between the Centre and the States and to recommend measures, including changes in the Constitutional provisions, to harmonise the relationship between the Centre and the States. While the Government has accepted and implemented several recommendations of the Administrative Reforms Commission, the recommendations of the Sarkaria Commission are under consideration in the Inter-State Council (ISC) which is trying to reach a consensus on various issues. Out of the 247 recommendations of the Sarkaria Commission, the Inter-State Council has taken a decision on 91 recommendations. These recommendations pertain to All India Services, Administrative Relations, Deployment of Union Armed Forces, Agriculture, Forest, Food, Civil Supplies, Mines & Minerals, Trade, Commerce, Mass Media etc. The re-activisation of ISC has opened a new chapter in the Centre-State relations and provided a useful forum for building a common national approach on various issues.

FORMULATION AND IMPLEMENTATION OF PLANS

In the sphere of planning, the balance of decision making process tilted more towards the Centre during the early years of planning. This was probably inevitable as there was lack of adequate experience in the formulation and the implementation of Plan programmes at the State level. No doubt, planning from the grass-roots level has been emphasised from the very beginning, but its implementation has been sporadic and tardy. Even so, the Planning Commission encouraged District Planning Committees (DPCs) and taluk level planning process from the Fourth Plan. Subsequently, a significant step was taken through the enactment of the 73rd and the 74th Amendments to the Constitution in 1993, which conferred a Constitutional status not only on the PRIs and the ULBs but also on the DPCs. It is now fully realised that the spirit of co-operative federalism should get reflected in the strong encouragement given to participative planning processes. In other words, the Central Government, the State Governments, the PRIs, the ULBs, and the non-Governmental organisations, the voluntary action groups and most of all, the people at the grass-roots have to be involved in the process of formulation and implementation of the Plans. It is realised that the process of democratic decentralisation can have true meaning only when sufficient autonomy and freedom is available to the States as well as to the PRIs and the ULBs in the formulation and the implementation of the Plans. At the same time, autonomy pre-supposes maturity in decision making and responsibility in the use of national resources. Decentralisation of the process of formulation and implementation should, therefore, ensure accountability.

Various steps have been taken in the past to improve the interaction with the States in the process of formulating and implementing the Five Year Plans. The National Development Council (NDC), which approves and reviews the Five Year Plans is chaired by the Prime Minister and includes Union Ministers and the Chief Ministers of all the States. In addition to this, periodic conferences of State Chief Ministers on major specific policy issues are held not only to elicit their opinion but also to take them into confidence in regard to the thinking of the Central Government. Over and above these, conferences of State Ministers of various development departments are also held to monitor the progress of Plan schemes and to find solutions to the problems faced by the State Governments in the course of implementing the Plan strategies, programmes and schemes. These conferences and meetings help resolve the day-to-day problems faced by the implementing agencies and the delivery system. At the administrative level, the secretariat of the Planning Commission interacts regularly with the administrative departments of the State Governments on all matters relating to the formulation and the implementation of the Five Year Plans. Thus, the regular and periodic interactions with the State Governments have helped in sorting out various policy issues and chalking out mutually agreed policies and programmes for achieving the goals of the Five Year Plans. There have also been frequent conferences and meetings of State Chief Ministers and Ministers at the zonal level to sort out the administrative problems relating to mobilisation of resources, implementation of Plan programmes and even maintaining law and order. In a vast country like ours, such periodical conferences and meetings of the State Chief Ministers and Ministers will go a long way in resolving certain unique problems relevant to the particular zones.

Similarly, at the State level, the State Governments should develop machanisms and forums for interacting with the elected representatives of the PRIs and the ULBs to resolve the issues relating to the formulation and the implementation of district and grass-root level development programmes. In some States like Karnataka, a State Development Council has been constituted, which is presided over by the Chief Minister and attended by the Presidents of Zilla Panchayats and Municipal Presidents. This is a replica of the NDC, which is worth emulation in other States also. There is also a need for a horizontal consultation process between the PRIs and the ULBs to learn from each other's experiences and to resolve day-to-day problems, if any. During the Ninth Five Year Plan such initiatives will be further encouraged and supported.

SHARE OF STATES IN PUBLIC SECTOR PLAN OUTLAY

An important area of concern in the sphere of planning is the decreasing share of the States in the overall public sector Plan outlay. The share of the States in the actual public sector expenditure is going down continuously from the Fifth Plan onwards, so much so that while the share of the States was about 50% in the Fifth Plan it came down to almost 37% in the Eighth Plan. Further, the share of the States in terms of total public sector expenditure has been lower than the originally envisaged outlay in successive Five Year Plans since the Sixth Plan. In the Eighth Plan, while the share of the States in the total public sector outlay was envisaged to be 41.46%, in actual realisation it is likely to be of the order of 37 percent. The main reason for the declining share of the States has been the shortfall in the mobilisation of resources by the State Governments. Besides others, the non achievement of the States' envisaged outlays has adverse implications for sectors like Agriculture, Education, Health and other Social and Basic Minimum Services.

The States' own resources (SOR) for their Eighth Plan were projected at Rs.1,01,485 crore. As against this, the expected realisation would be only Rs.70,335 crore, i.e., 69.3 percent of the projected SOR. The major shortfall was in respect of the Balance from Current Revenue (BCR) and internal contributions of public enterprises of the States. The BCR was projected at Rs.12,985 crore, while the expected realisation is (-) Rs.2,009 crore only, showing a shortfall of Rs.14,994 crore. An important reason for such a severe shortfall was the failure of many State Governments to raise the budgetary resources through additional revenue raising measures, increase in the non-Plan revenue expenditure on the target group oriented programmes, effect of the adoption of Central Pay scales and Central Dearness Allowances by many State Governments. The financially unviable operations of the State Electricity Boards, the State Road Transport Corporations and other State enterprises in most of the States had an adverse impact on the BCR of the State Governments, as these resulted in higher subsidies and grants to these enterprises, in addition to substantial loss of revenue in terms of interest and dividends, while States were forced to pay interest on their loans. The unwillingness of most of the States to revise the irrigation rates at least to meet the O&M expenditure was another reason for the shortfall in the States own resources. The most worrisome aspect was the increasing number of States providing free electricity to the farmers at a time when the country is facing a severe power shortage.

The internal resources of the State public sector enterprises was projected at Rs.4,000 crore for the Eighth Plan but the realisation is expected to be around (-) Rs.2,723 crore showing a deterioration in resources to the extent of Rs.6,723 crore. The shortfall in this was mainly due to the financially unviable operations of the State Electricity Boards and State Road Transport Corporations. Had the States implemented the agreed decision to levy a minimum of 50 paise tariff per unit of energy supplied to the agricultural sector and contained the T&D losses, thefts and arrears to a reasonable level, the SEBs would have been able to generate resources as committed in the Plan. The losses of the State Transport Corporations have been attributed mainly to social responsibilities undertaken by them to ply buses on uneconomic routes and providing concessions to a large number of weaker sections and students.

The Planning Commission has been emphasising that the States should plug the drain of resources in the operation of State level public enterprises, especially in the case of State Electricity Boards and State Road Transport Corporations. These include restructuring the State Electricity Boards, modernisation and other efficiency improvement as well as tariff revisions to make these Boards financially viable. The State Electricity Regulatory Commissions should also be set up without delay. The State Road Transport Corporations also should be made viable through proper fare revisions and improvements in the efficiency norms. They should be given support for strengthening their equity base. Further, the Planning Commission has emphasised the need for improving the Balance from Current Revenue (BCR) of the States through specific budgetary measures and containing the growth of non-Plan revenue expenditure. All these measures will improve the share of SOR which will enable the States to increase the size of their public sector Plan outlay. On its part the Government of India should help by providing a suitable macro economic framework. Efforts are also needed to build up a consensus on various issues like recovery of user charges, etc. to facilitate decisions at the State level.

TRANSFER OF CENTRALLY SPONSORED SCHEMES

Another issue which has been cropping up from time to time in the relations between the Centre and the States pertains to the Centrally Sponsored Schemes (CSS). The CSS are being implemented right from the beginning of the First Plan. When they were conceived, they were welcomed as they were intended to serve certain specific national goals. One classic example of CSS, which has made a far reaching impact, has been the introduction of HYV seed. This ushered in the Green Revolution in the country. But the number of CSS has grown over the years and they have made deep inroads even into the spheres falling within the domain of the States. There have been efforts from time to time for transferring the CSS to the States along with the corresponding funds, but not to much avail. A review undertaken by the Planning Commission showed that in 1995-96 there were 182 CSS under implementation with the Central funding of the order of Rs.16,000 crore. This level of Central funding is even more than the normal Plan assistance given to the States for the State Plans. In principle, the CSS should be confined to schemes of an inter-State character, matters impinging on national security, selected national priorities where Central supervision is essential for effective implementation and multi-State externally financed projects where the Central coordination is necessary for operational reasons. Except for such schemes, all other schemes should be transferred to the States along with the corresponding funds. A detailed exercise has been done by the Planning Commission to identify the schemes which could be transferred to the States in the light of the above approach.

A final decision in the matter would be taken after ascertaining the views of the State Governments/Central Ministries with the approval of the NDC.

INTER-SE ALLOCATION OF PLAN ASSISTANCE

Since 1969, the Central assistance for the State Plans has been distributed to the States as per the Gadgil Formula. Some marginal changes have been made in this formula from time to time, but the emphasis has always been on channelising larger Central assistance to the backward States. The formula was last amended in 1991-92. As per the existing formula, after setting apart the funds required for Externally Aided Projects (EAPs) and reasonable amounts for Special Area Programmes, 30% of the balance is earmarked for ten Special Category States. The balance amount is distributed among Non-Special Category States on the basis of following criteria and weightage :

Criteria		Weightage
Population (1971)	-	60%
Per Capita Income	-	25%
Performance Criteria	-	7.5%
Special Problems	-	7.5%

During the Plan discussions, various Chief Ministers have asked for a revision of this formula for allocation of Central assistance to the State Plans so as to change both the criteria and the relative weights. While the backward States have argued for higher weightage to be given to backwardness, (equity), the better-off States have argued that they should be rewarded for maintaining financial discipline and achieving better productivity, (efficiency). No formula can satisfy the demands of all the States. The formula for the allocation of Central assistance has to be such that it helps in a balanced regional development on the one hand and encourages financial discipline and better economic performance on the other. The solution to this contentious problem has to be found by a consensus. The views of the State Governments have been sought regarding the revision of the formula. The NDC will have to take a final view on this matter.

Nature of Central Assistance

Besides the issue of inter-se distribution of Central assistance among the States, the nature of Central assistance, that is, the grant-loan component, has also been raised from time to time. The States argue that when 30% of the grant component was introduced in 1969, the revenue component of the States' Plans was around 30 percent. Subsequently, it has gone on increasing, reaching upto 45 percent. They maintain that this justifies the increase in the grant component of Central assistance for State Plans under Gadgil Formula. Further, the States argue that among other reasons, the 70% loan component in the Central assistance for the States Plans has been a major reason for their increasing debt burden. They argue that their debt burden has been increasing so much so that in some cases there is net reverse flow of funds. They have, therefore, been asking for an increase in the proportion of the grant component in the Central assistance. The Special Category States have demanded Central assistance to be given as 100 per cent grant. The Sarkaria Commission had examined this issue and recommended different patterns of loan-grant assistance, depending upon the paying capacity of the States. The Finance Commissions in the past had also looked into the debt position of the States and recommended a scheme for general debt relief for all States, linked to fiscal performance and specific relief for States with high fiscal stress, the Special Category States and the States with debt problems warranting special attention. It has to be recognised that any change in the grant-loan component of the Central assistance has implications for the resource position of the Central Government. The NDC may have to decide on the issue of grant-loan proportion in the Central assistance while considering the question of inter-se distribution of the Central assistance among the States.

BASIC MINIMUM SERVICES (BMS) APPROACH

A special focus on meeting the minimum needs of the people has been given since the Fifth Five Year Plan. However, the minimum needs programme, which was formulated and implemented for this purpose, did not make much headway, partly because of allocation of inadequate funds. Because of the persisting gaps in BMS prevailing particularly in rural areas, it was decided to give a big push to this programme from 1996-97. Accordingly, the Central Government has made an allocation of a special Central assistance for seven Basic Minimum Services (BMS). Instead of being routed through the Central Ministries / Departments, this special assistance has been given to the States as additional Central assistance towards their Plan. With a view to ensuring that the special Central assistance for BMS is indeed an additionality, it has been emphasised to the States that this assistance should not be made a substitute for the normal provisions which were being made by the States for the BMS. By mutual consensus, it has also been agreed that the States should provide at least 15% of the special Central assistance for BMS as their share over and above their normal provisions for these schemes. While the States have been allowed flexibility in deciding their priorities among various components of Basic Minimum Services, the consensus at the Chief Ministers' Conference was to give the highest priority to the three services viz., Primary Health, Primary Education and Drinking Water for universal coverage. Only when the prescribed norms for these services have been achieved, the States could choose from other BMS for their attention. However, some States have deviated from the decisions of the Chief Ministers Conference and allocated huge funds for services other than the three which were to be given the highest priority. Another problem that has been encountered is that some States have failed to make adequate provisions for the BMS as per the agreed formula. Considering the difficulties of the Special Category States in mobilising resources, they have been exempted from providing 15 per cent contribution. To ensure additionality of Additional Central Assistance (ACA) over normal provisions of the States, the release of ACA will have to be linked to suitable provisions being made to BMS as per the decisions already taken. From the Annual Plan 1997-98 onwards the allocation of Central assistance for BMS has been made on the basis of identified gaps in providing these BMS. This process will continue during the Ninth Plan period and larger amounts would be made available to the States for meeting the requirements of BMS so that the basic minimum needs of the people are met by a target date. The approach adopted in respect of BMS has also been followed for the development of urban slums. The same model can be replicated in respect of other programmes, as may be found feasible, in future. Since there are backward regions within the States, it is expected that while making allocations, the States would also keep in view the requirements of the backward areas within the State so that the basic minimum needs of the people in these areas are also catered to.

PROCEDURE OF PLAN FORMULATION

The States should have full autonomy and freedom to draw up their Plans in accordance with the national priorities. There are, however, still some areas where the approval of Central agencies are required for the State projects, specially in the areas of power and irrigation projects as well as environmental clearances. Often, these clearances take long time leading to unnecessary delays and consequent time and cost overruns of the projects. The endeavour should be to delegate powers to the extent possible to the State Governments and also to ensure time-bound clearances in cases where the projects would need to be cleared at the Central level. The Planning Commission has already examined the position pertaining to the approval required for irrigation and power projects. As regards the power projects, investment approval was earlier necessary for all projects costing Rs.100 crore or more. After reviewing the position, a decision has been taken that power generation projects upto 100 MW (except where inter-State issues are involved) would not require clearances from the Planning Commission. Transmission schemes of only 220 KV and above are required to be referred to the Planning Commission. All renovation and modernisation schemes need not be referred to the Planning Commission. Similarly, guidelines have been issued that medium irrigation projects need not come for the approval of the Planning Commission (unless where inter-State angle is involved).

Efforts have been initiated to involve the State Governments more closely in the formulation of Five Year Plans. The occasion of the Annual Plan discussion for 1996-97 was utilised to ascertain the views of the State Governments relating to the objectives, approaches, priorities and strategies for the Ninth Plan. These views were taken into consideration while preparing the Approach Paper to the Ninth Plan which was subsequently endorsed by the NDC. The State Governments are also represented on various Steering Committees and Working Groups which are set up by the Planning Commission for various sectors to review the progress made in the previous Plans and to decide the prospects and strategies for different sectors. The reports of these Steering Committees and Working Groups are a valuable input to the Planning Commission in the formulation of Five Year Plans.

Considerable thought has been given to the issue as to how the State Governments can have greater autonomy in formulating their Plans within the national objectives and priorities as reflected in the Five Year Plan Document. The procedure for State Plan formulation was amended from the Annual Plan 1993-94. As per the procedure prior to 1993-94, the sectoral Working Groups set up in the Planning Commission and having representation from the State Governments used to project the requirements of different sectors which invariably totalled up to a much larger Plan size than was feasible keeping in view the availability of resources. The Plan size used to be fixed at the meetings between the Deputy Chairman, Planning Commission and the State Chief Ministers, after the sectoral Working Groups had projected their demands. Since the Plan size was not able to accommodate the projected requirements of different sectors, there used to be arbitrary cuts while fixing the final sectoral allocations. Since 1993-94, the procedure has been altered and the meeting between the Deputy Chairman, Planning Commission and the State Chief Ministers is organised to finalise the Plan size before working out the sectoral allocations within the Plan size so determined. Many a time, the

States show over-optimism regarding raising their own resources with a view to getting a higher Plan size fixed, as compared to the previous year even though the achievement in the previous year may be much less than the originally fixed Plan size. Fixation of unrealistic Plan size which does not materialise, has affected the credibility of the whole planning exercise. The Planning Commission would, therefore, in future indicate the level of Central assistance under the various heads to the State Governments and the States would decide their Plan size taking into account the Central assistance and their own resources and would seek legislative approval for both the Plan outlay as well as the additional measures for resource mobilisation. This system will encourage better accountability and commitment of the State Governments to achieve the Plan size as decided by them.

The interaction of the Planning Commission with the State Governments is at various levels in the process of formulation of Plans. An exercise is done by the financial resources Working Group to assess the likely availability of resources. Another exercise is done regarding the flow of assistance under externally aided projects. These interactions have proved very useful because while assessing the resources available under various heads, it is possible to assess the contribution of various items including the State public sector undertakings. The achievement in respect of tax and non tax revenues, vis-a-vis the potential, recovery of user charges, growth of non-Plan expenditure are also gone through in detail. These have not only provided crucial insight to the Planning Commission into various aspects of the functioning of the States' economy, but the interactions are useful for the States also because various deficiencies in the functioning of State PSUs or other sectors of the economy come to light. The position of different States in respect of various items facilitate comparisons and can encourage action in various areas.

The interactions with the State Chief Ministers at the level of Deputy Chairman have also proved to be mutually beneficial. The focus of these meetings has been changed from fixing the Plan size to discussing various sectoral issues, the performance of the States' economy, intra-State economic disparities, environmental status and remedial steps which may be necessary to improve the shortcomings. Any problems which the State may have with the

Central Government are also projected at these meetings. Model practices adopted in various States in different schemes/projects are also shared at these meetings so as to learn useful lessons.

Another level of interaction with the State Governments is through the mechanism of the Working Groups, which are set up in the Planning Commission to decide the sectoral allocations within the overall Plan size fixed at the level of Deputy Chairman and Chief Ministers. These Working Groups have, in the past, not only been utilised to finalise the sectoral allocations, but have also provided an important forum of mutual exchange of views regarding the performance in the previous years, the strategies for development in future and any problem areas. Useful data is also exchanged at these Working Group meetings pertaining to various aspects. These Working Group discussions would, therefore, continue, albeit the number of suchWorking Groups may be reduced to a minimum.

It is essential for the Planning Commission to know the detailed sectoral Plans of the States at least in respect of the sectors crucial for the national economy like power, irrigation, communication, other infrastructural sectors, etc. so that the growth and strategies at the national level can be worked out. While working out the detailed Five Year Sectoral Plans, the policy perspective for private sector participation and the investment expected have to be considered. Once the Five Year Plan outlays have been finalised, the annual phasing of sectoral /sub sectoral allocations can be decided by the State Governments keeping in view the approved Five Year Plan outlays, national priorities and their own needs. A detailed examination of these sectoral allocations at the Planning Commission level is not necessary. The Annual Plan documents prepared by the States would be examined in the Planning Commission to ensure their broad adherence to the Five Year Plans.

EARMARKING OF OUTLAYS

An area which has often been mentioned as affecting the autonomy of States in allocating resources among different sectors within the overall Plan size pertains to earmarking of outlays for particular sectors/schemes/projects. Upto 1969, the Central assistance used to be given for individual schemes/projects. On the recommendation of the NDC it was, however, decided that the Central assistance may be given in block form. However, to ensure compliance with the national priorities the system of earmarking of outlays for crucial sectors/schemes was initiated. Presently, the earmarking is restricted to 50% of the Plan outlay for Non-Special Category States while for Special Category States it can be higher. Under the existing guidelines the earmarking is being followed for the following

i. Selected ongoing projects for power and irrigation;

ii. Externally aided projects;

iii. Schemes in one or two sectors which are of importance to the State's economy And fall priority areas of Five Year Plans;

iv. Basic Minimum Services.

The areas and the actual projects to be earmarked are settled with the State Government officials during the Working Group discussions. The earmarking has helped in ensuring minimum provisions for important schemes/projects in accordance with the national priorities. The system would continue even though a flexible approach would be adopted and earmarking would be restricted to the minimum extent necessary.

REVIEW AND MONITORING

Under the present system of review and monitoring, a brief general review of the performance of the States, specially in financial terms, is made at the meetings at the level of Deputy Chairman and the Chief Ministers. A more detailed exercise is done through the mechanism of Working Groups set up under the auspices of Planning Commission. The Working Groups set up for major sectors meet to discuss the performance of the previous year and the proposals of the following year including the need for earmarking for particular schemes / sectors on one day for each State. However, this exercise is done once a year and there is no mechanism for getting regular feedback from the States regarding the implementation of the various Plan schemes.

In order to be effective, the review and monitoring of the State Plans has to be a participative exercise with the State Governments. Further, these detailed exercises for review and monitoring may not be held only at Delhi, but in various State capitals to the extent feasible. Action in this direction has already been initiated and meetings at the level of Deputy Chairman in the State capitals have already been started. More such meetings would be organised in the State capitals.

Equity-Promoting, Holistic and Coordinating Role of Central Planning

While the question of greater autonomy to the States in the process of development planning is considered, the holistic, equity-promoting and coordinating role of central planning cannot be over-looked. The role of central planning has to be seen not as a constraint to the State's autonomy but as a necessary complement to decentralised, flexible and cooperative mode of development planning. The equity-promoting role demands that greater efforts are made to remove the gaps in the provision of BMS, the level of development of rural agricultural hinterland and infrastructure so that no region or sub-region and no group or groups of people remain deprived of the fruits of development and every region and all people at least reach a minimum standard of living. The equity-promoting role of central planning assumes added importance in the wake of the emerging policy environment in the country as well as globally. With the opening up of the economy and removal of controls, the market forces have been unleashed, which may have a tendency to exacerbate the disparities among groups of people and regions which can be further widened because of the operation of large global players. As the economy gets integrated more and more with the global economy, the Centre may be required to play a stronger equity-promoting role and to secure sufficient space for all the federal units to work out their own strategies of development without being overtaken by global regimes or forces. The Centre would also be required to ensure suitable macro economic policy framework for the growth of economy for meeting the aspirations of the people. A judicious balance on the part of the Centre between decentralisation and autonomy on the one hand and intervention to protect the interests of the weak and to provide space for autonomous decision making on the other is required to sustain the faith of the States in the philosophy of co-operative federalism.

Mobilisation and Sharing of Resources

The sphere of financial relations between the Centre and the States has been the most difficult one and despite continuous efforts, unanimity has not been reached on various issues involved. The issues, which have a bearing on resource availability, need careful analysis and discussion. The Sarkaria Commission had paid enough attention to the issue of financial relations between the Centre and the States and had made a large number of recommendations in this regard. These are being considered by the Inter-State Council and its Standing Committee.

Essentially, the problem of resource availability, which affects both the Centre and the States, can be overcome only if there are greater efforts to tap the resource potential both at the level of Central Government as well as at the State Governments. Unless the resource availability of the Central Government improves, it may not be possible to channelise a larger quantum of Central assistance or to increase the grant component to the States. The Government of India has recently agreed to a share of 29% to the States in the total gross revenues from all the Central taxes with effect from April, 1996. This will increase the flow of funds in absolute terms on non-Plan account. With this increased share of the States in the total devolution, the States would be in a position to benefit from the revenue buoyancy from all the taxes levied by the Centre. While the share of 29% of the net yield from all Central taxes has not satisfied all the States, the decision itself marks a watershed in the approach to the sharing of tax revenues of the Centre, because the sharing which was so far limited to two taxes, viz., Excise and Income Tax, will now include all Central taxes. In other words, the divisible pool of Central taxes has been broadened by implementing the Alternative Scheme of Devolution recommended by the Tenth Finance Commission.

Notwithstanding such a widening of the divisible pool of Central taxes, the financial needs of the States will increase during the Ninth Plan because of the large gaps which remain in the field of BMS and infrastructure. Therefore, the States need to raise more of their own resources to meet such increased outlay. In order to improve the resources of the States and to enable them to raise the required level of resources for Plan purposes, the Centre has to play some supportive role in certain specific areas. These areas relate to statutory devolution of funds from the Centre to the States through Finance Commission, discretionary grants and loans through Planning Commission, revision of royalty rates once in two years and application of these rates on ad valorem basis, helping the States to move towards a uniform tax structure and Modvat with the possibility of imposing Consignment Tax in the interim period. The Central Government has already constituted the Central Electricity Regulatory Commission under the Electricity Regulatory Commission Act, 1998. Some of the States have also set up their own State Electricity Regulatory Commissions which, inter alia, provide for unbundling and corporatisation of State Electricity Boards, fixation of electricity tariff and introducing competition. Many other States are in the process of setting up their own Electricity Regulatory Commissions in due course. All these steps are expected to improve the financial health of the State Electricity Boards which would result in increased size of State Sector Plan Outlay.

INTER-STATE RELATIONS

Harmonious and cooperative relations between different States are as important as that between the Centre and the States for the healthy functioning of our federation. Various problems have been cropping up in inter-State relations from time to time. On the one hand, there are problems like increasing competition among the States and on the other, there are disputes over sharing of river water etc.

With the dismantling of controls and greater freedom for the location of industries, the competition among the States for attracting industries and other economic activities by offering incentives has intensified. While a healthy competition among the States for providing better and efficient services is to be welcomed, the practice of granting tax rebates and subsidies need to be seen in the right perspective of whether they lead to national welfare. Instances are not lacking where the States have joined a sort of a race in granting sales tax and other rebates for attracting new industrial units. This has affected the resource position of the concerned States without commensurate benefits, because in so far as the same concessions are allowed by a number of States, these concessions do not remain the main guiding factor for the location of an industry and other important considerations like efficiency of services and infrastructure facilities may become more important.

Similarly, often there is a tendency on the part of the States to allow tariff concessions to various sections. Power tariff is a case in point and many States have been supplying power to various sectors at a price much lower than the cost of generation. A national consensus had evolved at the Chief Minister's Conference for fixing the minimum tariff for agriculture and to move over to a regime of fixing tariffs for recovering the costs. Some States have, however, gone against this consensus and allowed free electricity to agriculture sector. This has not only put the financial condition of the State Electricity Boards in a precarious position but also created problems for the neighbouring States where vociferous demands have been made for similar concessions. Populist measures have a tendency to spread because once they are granted by one State, pressure builds up in the neighbouring States for allowing the same concessions. Similar is the position regarding the administrative charges for other services like irrigation and water. Despite various Commissions recommending minimum rates for such services, so that at least O & M costs are recovered, many States have not implemented these recommendations, depriving them of an important source of revenue and putting their financial position under great strain. It is essential to have a national consensus regarding such issues like harmonisation of tax structure, minimum tariff for certain services, cap on the level of subsidies and facilitation of inter-State trade flows. It is necessary that all the States should fall in line for implementing mutually beneficial policies.

Another important problem that keeps cropping up time and again is in respect of share in natural resources like river waters which has given rise to severe tensions and prevented optimal utilisation of such resources. There is a need for evolving a national policy on this issue and for putting in place a suitable mechanism for resolving such disputes.

10

Special Area Programmes

The objectives of the Ninth Plan have been spelt out in the Approach to the Ninth Five Year Plan document adopted by the National Development Council. "Growth with Social Justice and Equity" is proposed to be achieved with the objectives outlined in the Approach Paper. Some specific areas from within the broad objectives of the Plan as laid down by the NDC have been selected for special focus. For these areas, Special Action Plans (SAPs) have been evolved in order to provide actionable, time-bound targets with adequate resources. Broadly, the SAPs cover specific aspects of social and physical infrastructure, agriculture, information technology and water policy.

Volume-I of the Ninth Plan document covers not only the macro and economy-wide issues but also reflects on the important general aspects of the sector specific issues, particularly the approaches to policy formulation. The main thematic areas covered in Volume-I, inter alia, include Economic infrastructure, energy, agricultural development and food security, industry and commerce, and the financial sector.

Volume-II gives perspectives on how the sectoral programmes fit into realisation of planning themes enumerated in Volume-I. The sectoral chapters deal with the trade-off in objectives both for the perspective period and during the Ninth Plan, commonality in approach, policy framework, institutional structure, delivery mechanism, possibility of synergy and convergence and major requirements i.e. labour, input and technology.

In individual sectoral chapters, policies and programmes during the Eighth Plan period have been reviewed, shortcomings identified and new policy framework suggested to overcome the shortcomings and ensure sustainability of the development process not only in economic terms but also in terms of social and environmental factors. Sectorwise overview is discussed in the succeeding paragraphs:

AGRICULTURE, FOOD SECURITY AND IRRIGATION

The Strategy of agricultural development would be centred around achieving the objectives of sustainability of employment generation, food and nutrition security, equity and poverty alleviation. Efforts will be made to achieve a growth rate of 4.5 per cent per annum in agricultural output in order to make a significant impact on overall growth and poverty. Regionally differentiated strategies will be followed to realise the full potential of growth in every region. The emphasis will be on raising the capabilities of small peasants and promoting sustainable agricultural systems, while at the same time conserving and maximising the value from scarce resources, water and land. Infrastructure development will be given the highest importance. Emphasis will be laid on minor irrigation by harnessing ground water resources. Timely and adequate availability of inputs will receive special attention. The regional programmes will be formulated in such a manner as to ensure provision of inputs to the farmer, particularly in the remote, hilly, backward and tribal areas. Agricultural credit is a crucial input and it will receive special attention. The programmes relating to land reforms would be strengthened to raise agricultural growth and help the poor. Efforts will be made to increase public investment during the Plan period. In every district, the Rural Infrastructure Development Fund (RIDP) would be used to promote projects which encourage organisations of groups of small farmers, artisans and landless labourers for skill up-gradation, processing, transport infrastructure, quality improvement etc. Support to agricultural research will be enhanced and emphasis will be placed on bio-technology, microbiology, genetic improvement of crops including hybrid technology, genetic up-gradation of animal resources, improvement of fish genetic stock and post-harvest technology, etc. Efforts will be made to accelerate the growth rates of allied sectors such as horticulture, including fruits and vegetables, fisheries, livestock and dairy. Agricultural exports will receive special attention as these have a lot of potential for increasing farm incomes and employment, besides earning foreign exchange. Co-operatives will be strengthened. Greater participation of women in agriculture than at present will be encouraged. Linkages with markets will be strengthened and agro-processing and agro-industries will be encouraged.

Self-sufficiency in food-grains for the country as a whole does not necessarily imply food and nutrition security for all. Making food available through the PDS at affordable prices has been a key element of our food security system. However, untargeted PDS has resulted in a steady rise of budgetary food subsidy. During the Ninth Plan subsidised food grains is proposed to be targeted only to people below the poverty line so as to ensure that budgetary subsidies reach the needy and become sustainable. In addition to reorienting PDS, food supplementation programmes for the identified vulnerable groups such as women and children (such as ICDS, Mid-day Meal, Vitamins and Iodine deficiency programmes) are proposed to be extended and strengthened.

INDUSTRY & MINERALS

The thrust of the new industrial policy announced in July 1991 has been on substantial reduction in the scope of industrial licensing, simplification of procedures, rules and regulations, reforms in the Monopolies and Restrictive Trade Practices(MRTP) Act, reduction of areas reserved exclusively for the public sector, disinvestment of equity of selected public sector enterprises (PSEs), raising the limit of foreign equity participation in the domestic industrial undertakings, liberalisation of trade and exchange rate policies, rationalisation of duties, etc. with a view to promoting growth and increasing efficiency and international competitiveness. Steps were also taken to bring public sector enterprises within the ambit of Board of Industrial and Financial Reconstruction (BIFR), providing a social security net through National Renewal Fund (NRF), industrialisation of backward areas through growth centres scheme, opening up of mining sector to private sector, strengthening of technological capacity etc. Industrial growth rate during the Eighth Plan was, however, lower than that achieved in the Seventh

Plan. One of the major constraints faced by the industrial sector is inadequate availability of infrastructural support, which not only affected domestic production but exports as well. However, the industry has responded well to the opening up of the economy and has realised the importance of competitiveness with the result that due emphasis is being given to modernisation, up-gradation, economies of scale, quality, research and development, etc. The rate of foreign direct investment has also started picking up in response to the improved policy environment.

The Ninth Plan envisages an industrial growth rate of 8.50% per annum and export growth of 11.8% per annum. For achieving this growth, special measures have been suggested to ensure adequate availability and requisite quality of infrastructure and creating conditions conducive for unhindered growth of such industries which can produce products at internationally competitive prices. Internal aberrations in policies are proposed to be removed and special measures envisaged to promote development of industries in backward areas. Special emphasis has been given to the industrial and economic development of the North Eastern Region by evolving a special package which, inter alia, includes changes in the funding pattern of growth centres and integrated infrastructure development centres, extension of transport subsidy scheme, strengthening of institutions concerned with entrepreneurship and human resources development etc., various physical concessions and incentives and specific measures for development of industrial sub-sectors like agriculture, handicrafts, handloom etc. It is also proposed to review the working of BIFR and bring about necessary changes to make it an effective instrument of reviving sick industrial units. The scheme of National Renewal Fund is also proposed to be recast through appropriate modifications to make it more effective and achieve the objectives of providing a social safety net as originally envisaged. The retention price-cum-subsidy scheme for fertilisers and dual pricing scheme for sugar is also proposed to be reviewed and necessary changes will be made with a view to ensuring healthy development of these industries.

The small scale sector has shown considerable resilience and in-built strength and growth rate of this sector has been about two to three percentage points higher than that of large and medium industries. Due emphasis would be given for making available adequate credit to the sector, and promote production and productivity through technological up-gradation.

In the unorganised sector, a cluster approach will be adopted for provision of training, up-gradation of skills and improvement in tool kits, equipment and production techniques to increase production, productivity and income levels of artisans, craft-persons, weavers, spinners and workers etc.

Energy

The major thrust of the on-going reform process in the Energy Sector has been to make it commercially viable and also to attract private sector participation. The Ninth Plan lays emphasis on bringing about a commercial outlook among the PSUs, attracting private sector participation in the development of energy sector, encouraging a competitive environment not only between public and private sectors, but between public sector units, need for regulatory agencies for fixation of tariff, conservation of resources, safeguarding the interest of consumers and protecting the environment. Energy-economy interaction, viz. the demand for energy in the economy, the sources from which this demand is met, the changes in the pattern of energy consumption from non-commercial to commercial, changes in the composition of commercial energy use, energy-GDP elasticity along with some policy measures have been discussed. New policies in the Ninth Plan on getting the private sector to invest in the electric power generation, transmission and distribution, in coal mining, in petroleum exploration, production and arranging of supply of liquefied natural gas have been considered. Restructuring of the electricity supply system to make it commercially viable, bankable and professional, rapid increase in oil and coal production by deregulating the industry in a short period of time are some of the aspects which will receive due consideration.

Transport

The country's transport system which comprises of rail, roads, sea port and airports is facing capacity saturation. Inadequacies and imbalances in transport threaten to constrain economic growth and the quality of life in both urban and rural areas. A large number of villages lack reliable all-weather connection with nearby markets and towns. Areas like North East and Jammu & Kashmir have remained physically and emotionally isolated because the transport system has not linked them adequately with the rest of the country. Environment friendly and socially cost-effective means of transport like coastal shipping, inland water transport and non-mechanised transport, etc. also have remained undeveloped. Distortions in the inter modal mix of transport, environment and energy linkages, safety and technology up-gradation have been examined. The diverse issues facing the transport sector require a comprehensive policy package. Strengthening of the Indian railways in its reach and capacity so that it effectively links the distant parts of the country, helps to develop economic potential of the backward areas and carries the bulk of the nation's long or medium haul traffic, is necessary. Similarly road networks need to be expanded and strengthened. There is also a need to modernise our seaports and airports with a view to augmenting their capacity and making them of international standard.

Policy options, each backed by adequate investment and complemented by suitable policy changes in other sectors, have been discussed and emerging issues and strategies visualised for development have been elaborated. The pricing policy in future would have to be based on full recovery of cost. Subsidies/ concessions will be direct through General Budget or sectoral budget so that the health of the transport enterprises is not impaired. Attempts will be made to mobilise resource from user charges in various ways. Non-tariff measures will also be taken for resource generation. Revenue mobilisation through tariff and non-tariff measures, stress on productivity of human and material assets and cost cutting will be pursued vigorously in the public sector transport enterprises.

Measures will be taken for increasing the involvement of private capital in the expansion and strengthening of infrastructure in railways, road, shipping and airports. The private participation can take many forms like full or joint ownership management contract, leasing, concessions like BOT etc.

COMMUNICATIONS, BROADCASTING AND INFORMATION TECHNOLOGY

Information Technology (IT) is fast emerging as the technological infrastructure for global integration and rapid development of all sectors of economy. IT broadly includes all sub-sectors dealing with the generation, transmission and utilisation of information like telecommunications, computers, consumer electronics, media infrastructure etc. Recognising the impressive growth the country has achieved in IT since mid-eighties, its immense potential for future growth and its importance as an agent of transformation of every facet of human life, a high priority has been accorded to the development of this sector. The Ninth Plan would aim to make India a global IT super power and a front runner in the information revolution. To help devise an appropriate policy framework towards achieving this objective, a National Task Force on Information Technology and Software Development has been set up with Deputy Chairman, Planning Commission as its Chairperson. Expansion of the telecom network and its transformation into a modern and efficient system would be the two thrust areas during the Ninth Plan. Universal coverage or telephones on demand, universal and easy accessibility, world standard services to consumers at affordable prices, demand based provision of existing value added services and introduction of new services would be the major objectives for the Ninth Plan. To achieve the objective of providing telephones on demand, 237 lakh new telephone connections are envisaged to be provided.

ENVIRONMENT & FOREST

The Agenda 21 of the Rio Summit calls for integration of environmental aspects with development aspirations. One of the objectives of the Ninth Five Year Plan is to ensure environmental sustainability through social mobilisation and participation of people at all levels. It is also based on the belief that the principal task of planning in a federal structure is to evolve a clear vision and commitment to the national objectives and development strategy by both central, State and local governments. Therefore, emphasis is being placed on reorienting the policies rather than on direct intervention so as to send proper signals and induce economic development. The strategy in the Ninth Plan for the environment sector has been drawn up in accordance with the development needs of the nation. The measures required to protect the environment will be taken in such a way as to achieve sustainable development. The strategy recognises the symbiotic relationship between tribals and the forests, and gives a special focus to the scheduled castes, the tribals and the weaker sections living in and around the forests. A number of enabling conditions have already been created for harmonising economic growth and environmental conservation. These include the macroeconomic compatibility, the implementation of the 73rd and 74th Constitutional amendments and the measures being undertaken in the implementing ministries.

HUMAN AND SOCIAL DEVELOPMENT

Human Development and improvement in the quality of life are the ultimate objectives of planning. This is to be achieved through policies and programmes aimed at promotion of both equity and excellence. Benefits of national economic programmes reach the different segments of the population through different channels and at different rates. Economic growth improves employment opportunities and employment improves income and purchasing power. But the market mechanism may not improve access to available facilities or fully meet the essential needs of the population with inadequate purchasing power. Social sector planning therefore ensures that appropriate policies and programmes are formulated and adequate investment provided by the State so that poor and vulnerable segments of the population can access essential commodities and facilities based on their needs and not on the ability to pay.

Despite the fact that there has been a decline in the incidence of poverty over the past two decades, millions continue to live below the poverty line, a large proportion of whom reside in rural areas. Therefore, specifically designed anti-poverty programmes for generation of both self-employment and wage-employment will continue in the Ninth Plan. These would, however, be rationalised and re-designed in order to make them more effective as instruments of poverty alleviation. Under the self-employment programme of Integrated Rural Development Programme (IRDP) there would be a progressive shift from the individual beneficiary approach to the group and/or cluster approach. A holistic approach would be adopted with an integration of the existing sub-schemes of Training of Rural Youth for Self Employment (TRYSEM), Development of Women and Children in Rural Areas (DWCRA), Supply of Improved Toolkits to Rural Artisans (SITRA), and Ganga Kalyan Yojana (GKY) in the IRDP. Jawahar Rozgar Yojana (JRY) will be confined to the creation of rural infrastructure according to the felt needs of the people at the village level through panchayats. However, at the block and district level, the Employment Assurance Scheme (EAS) would be the single wage-employment programme. Efforts would be made to bring about a greater integration between the poverty alleviation programmes and the various sectoral programmes as well as the area development programmes within the umbrella of the Panchayati Raj Institutions (PRIs).

In the area of basic minimum services, the Ninth Plan has placed greater emphasis on primary health care, primary education and provision of safe drinking water and shelter.

Over the last five decades a massive health care infrastructure has been built up in urban and rural areas. National programmes to combat major health problems have been evolved and implemented through this infrastructure. These have resulted in a steep fall in mortality.

However there has not been any reduction in the communicable diseases and nutrition related morbidity. There has been a progressive increase in the non-communicable disease burden, occupational and environmental health related problems. Focus during the Ninth Plan will be to provide integrated preventive, promotive, curative and rehabilitative services for communicable, non-communicable and nutrition related health problems, through appropriate strengthening of the existing health care institutions and ensuring that they are optimally utilised. Efforts will be made to achieve substantial reduction in morbidity and mortality rates by taking advantage of the ongoing demographic transition and progressive increase in the population in the 15-59 age group.

Over the last five decades there has been a steep decline in severe grades of under-nutrition and related health problem. Currently major nutritional problems of public health importance are (a) chronic energy deficiency (CED) especially in pregnant women and pre-school children; (b) obesity and consequent increase in non-communicable diseases especially among urban affluent segments of population; (c) micro-nutrient deficiencies of iron, iodine and vitamin A and associated health hazards. During the Ninth Plan, efforts will be made to achieve substantial reduction in CED and its health consequences through universalisation of ICDS, screening of pregnant and lactating women, growth monitoring and better targeting of food supplements to those with CED, close monitoring of persons receiving food supplements; close inter-sectoral co-ordination to ensure early detection and management of health problem leading to or associated with under-nutrition. Prevention, early detection and effective management of micro-nutrient deficiencies and associated health hazards will receive due attention.

The technological advances and improved quality and coverage of health care resulted in a rapid fall in Crude Death Rate (CDR) from 25.1 in 1951 to 9.8 in 1991. In contrast, the reduction in Crude Birth Rate (CBR) has been less steep, declining from 40.8 in 1951 to 29.5 in 1991. As a result, the annual exponential population growth rate has been over 2% in the last three decades. During the Eighth Plan period the decline in CBR has been steeper than that in the (CDR) and consequently, the annual population growth rate has been around 1.9% during 1991-95. Reduction in the population growth rate has been recognised as one of the priority objectives during the Ninth Plan period. This will be achieved by meeting all the felt-needs for contraception; and reducing the infant and maternal morbidity and mortality so that there is a reduction in the desired level of fertility. The efforts will be made to minimise the existing disparities between states by providing resources to fill the crucial gaps in infrastructure and manpower and improving the operational efficiency of health system; improving the access and quality of reproductive and child health services to enable the increasingly aware and literate families to attain their reproductive goals in harmony with the national goals and accelerating the rate of reduction in population growth rate to achieve rapid population stabilisation.

Manpower development to provide adequate labour force of appropriate skills and quality to different sectors essential for rapid socio-economic development and elimination of the mismatch between skills required and skills available has been a major focus of human resource development activities during the last fifty years. Employment generation in all the productive sectors is one of the basic objectives. In this context, providing enabling environment for self employment has received special attention both in urban and rural areas. Objective is also to eliminate bonded labour, employment of children and women in hazardous industries, and minimise occupational health hazards. During the Ninth Plan period, elimination of such undesirable practices as child labour, bonded labour, ensuring workers' safety and social security, looking after labour welfare and providing of the necessary support measures for sorting out problem relating to employment of both men and women workers in different sectors will receive priority attention. It is also envisaged that the employment exchanges will be reoriented so that they become the source of labour related information, employment opportunities and provide counselling and guidance to employment seekers.

The Ninth Plan treats Education as the most crucial investment in human development with objective of eradicating illiteracy by 2005. Special Action Plan has identified the expansion and improvement of social infrastructure in education as a critical area. An overriding priority will be given for providing access to schooling to children in the age group 6-11 years. A special thrust will be on girls' education by providing free education up to college level including professional courses. Vocational education at the secondary and under-graduate levels will be expanded and restructured so as to have strong linkages with industry and improve employability. The intake capacity of the IITs, other reputed engineering institutions and IIMs will be doubled, particularly in high demand areas like software engineering and information technology. PRIs will be empowered to serve as the nucleus in programme implementation. Non-governmental organisations will be encouraged to supplement the governmental efforts, private sector will also be facilitated to grow particularly in higher and technical education. Care will, however, be taken to ensure that the country does not lag behind in terms of creative artists and scientists. Accordingly, education in basic sciences and areas of fundamental research will be promoted and strengthened.

To ensure the well-being of the disadvantaged sections of the society, there has been a definite shift in the approach from 'welfare' to 'development' and to 'empowerment' over the last four and a half developmental decades. In line with this new approach, all the welfare and developmental measures have been directed towards empowering the Socially Disadvantaged Groups, such as women, Scheduled Castes, Scheduled Tribes, Other Backward Classes and Minorities, as the agents of social and economic change and development. While in the case of women, efforts are now being made to make them economically independent and self-reliant through various employment and income-generating activities, the main thrust in respect of children is to ensure their survival, protection and development with a special focus on the girl child and the adolescent girl. In the field of social welfare, the special focus will be now on empowering the persons with disabilities; reforming the deviants like juvenile delinquents/ vagrants, alcoholics and drug addicts; and caring for other disadvantaged, viz. street children, older persons and destitutes. The main strength of these efforts is derived from various legislations enacted for protecting the interests of these groups. Towards empowering the Socially Disadvantaged Groups, a three-pronged strategy of Social Empowerment, Economic Empowerment and Social Justice, will be adopted in the Ninth Plan.

HOUSING, URBAN DEVELOPMENT AND WATER SUPPLY

The key urban concern is the growing gap between demand and supply of basic services. While there has been a steady growth in the housing stock, infrastructure and basic services, the gaps between demand and supply have been growing even in terms of conservative norms. This gap has a deteriorating impact on the urban environment, inadequacy of urban planning, urban poverty and degradation. With a view to achieving the goal of 'Health for all', the Government is committed to provide drinking water to every settlement in the rural and urban areas within next five years. It has also been decided to ensure that sanitation facilities are improved and expanded rapidly. A new National Housing and Habitat Policy has been evolved which aims at providing 'Housing for All'. Towards this end the Government will facilitate construction of 20 lakh additional housing units annually in urban and rural areas. To achieve these objectives Special Action Plans have been prepared.

The rural hinterland has played a critical role in sustaining urbanisation. This is reflected in the indicators of sources of primary inputs, competitively priced labour for urban economic activities, primary funds as reflected in comparative urban and rural credit-deposit ratios and market for urban pockets. But the unending migration of the rural poor to the urban areas may have a destabilising effect of urbanisation and its sustainability. Income and employment opportunities will have to rise in rural areas through both the farm and non-farm sectors. The rural-urban continuum would be strengthened so that the gaps between rural and urban life styles are reduced. Effective urban strategies and programmes cannot be developed in isolation of those living in rural areas. The Ninth Plan will take cognisance of this ground realities, particularly in respect of the three critical components viz. drinking water, sanitation and housing.

Science and Technology

The Science and Technology policy and approach for the Ninth Plan reflects the reality of the present day world in which nations progress along their own chosen path but in a much more closely inter-connected and inter-dependent manner. Science and Technology continues to remain the main focal point for exploring new horizons and new vistas, economic prosperity and meeting the economic, industrial, trade and societal challenges. Policy initiatives have been suggested to ensure the benefits emerging from the S&T reach all sections of the community including the weaker sections of the society. Since a strong science base is a pre-requisite for achieving technological competitiveness, efforts will be continued to build and maintain the same. Scientists with exceptional capabilities would be nurtured and supported fully by offering them within the country facilities comparable with international standards, by creating more centres of excellence in institutions of higher learning for supply of future S&T manpower and by utilising the existing infrastructure in terms of facilities and manpower more efficiently. Research programmes, particularly in some of the chosen fields of agriculture, export and industry would be taken up on in a mission mode through appropriate restructuring and reorientation of many of the scientific institutions and laboratories. The emphasis will be on clean and eco-friendly technologies. The major focus of the S&T programmes will be to encourage and strengthen interaction among R&D institutions and the users. Development of core strength and concentration on areas where competitive strength can be built so that the technological competitiveness can be converted into commercial strength are envisaged. In the light of the international control regime, greater awareness would be created among the scientists and technologists regarding the patents and IPR related issues for protecting the interest of the country. Science and technology activities in the States and UTs would be geared up to take up location specific R&D programmes for S&T inputs in the key sectors of the socio-economic development through promotion of joint innovative programmes with industry and NGOs.

The importance of developing S&T in a major way has been recognised since independence. The whole hearted support provided to science and technology since then has resulted in many accomplishments in a wide variety of disciplines. Support to basic research has been receiving a rather high priority during the earlier plans. Though this may continue to a considerable extent, a proper balance would be maintained between fundamental research and applied research in scientific fields. Recent developments have brought home the need to accord high priority to technology related areas, particularly the process technologies, which may be characterised as core technologies, which need to be strengthened with particular emphasis on ensuring partnership with the concerned socio-economic activities and industry wherever possible. In this process the Indian industry and the users of technology have a crucial role to play in generation of technology in order to sustain a competitive technological edge.

11

Poverty Alleviation Programme

POVERTY ALLEVIATION IN RURAL INDIA : PROGRAMMES AND STRATEGY

Poverty eradication is one of the major objectives of planned development. The magnitude of the problem is still quite staggering. Thirty six per cent of the Indian population was below poverty line (BPL) in 1993-94, the latest year for which the data are available and the absolute number of poor was 320 million, out of which 244 million (37 per cent of the rural population) lived in rural areas. The incidence of poverty declined from 54.9 per cent in 1973-74 to 36 per cent in 1993-94. But the absolute number of poor did not decline much over this period of 20 years. There were 321 million poor in 1973-74 and 320 million in 1993-94; in the rural areas the corresponding numbers were 261 million and 244 million.

The main determinants of poverty are (i) lack of income and purchasing power attributable to lack of productive employment and considerable underemployment and not to lack of employment per se; (ii) a continuous increase in the price of food, especially foodgrains which account for 70-80 per cent of the consumption basket; and (iii) inadequacy of social infrastructure, affecting the quality of life of the people and their employability.

Economic growth is important. Economic growth creates more resources and has the potential of creating more space for the involvement of the poor. But the involvement of the poor depends on the sources of growth and the nature of growth. If the growth is sourced upon those sectors of the economy or those activities which have a natural tendency to involve the poor in their expansion, such growth helps poverty eradication. Therefore, it is important to source a large part of economic growth in agriculture, in rural non-agricultural activities and in productive expansion of the informal sector which all have high employment elasticities, as well as in an export strategy based on labour intensive exports.

The Government recognises that high growth of incomes is by itself not enough to improve the quality of life of the poor. Unless all the citizens of the country, and most particularly the poor, have certain basic minimum services, their living conditions cannot improve. These minimum services include among other things literacy education, primary health care, safe drinking water and nutritional security. The Government had convened a meeting of Chief Ministers to identify such basic minimum services and a list of seven services had unanimously been agreed upon. These seven services are safe drinking water, primary health facilities, universal primary education, nutrition to school and pre-school children, shelter for the poor, road connectivity for all villages and habitations, and the Public Distribution System (PDS) with a focus on the poor. The Ninth Plan lays special emphasis on these seven basic minimum services and will make all efforts to achieve a minimum level of satisfaction in providing these in partnership with the State Governments and the Panchayati Raj Institutions (PRIs).

Direct poverty alleviation programmes are important and will continue on an expanded scale in the Ninth Plan. But these programmes would be oriented towards strengthening the productive potential of the economy and providing more opportunities for involving the poor in the economic process. Broadly, there would be schemes for income generation through supplementary employment, for the welfare of the poor in rural/urban areas and for a targeted PDS system to ensure that the poor have access to foodgrains at prices they can afford. In this chapter, both rural and urban poverty alleviation programmes besides the Targeted Public Distribution System (TPDS) will be discussed in some detail. Poverty can effectively be eradicated only when the poor start contributing to the growth by their active involvement in the growth process. Implementation of the programmes should be increasingly based on approaches and methods which involve the poor themselves in the process of poverty eradication and economic growth. This is possible through a process of social mobilisation, encouraging participatory approaches and institutions and empowerment of the poor. In this the PRIs, the voluntary organisations and community based Self-Help Groups will be more closely involved.

RURAL POVERTY ALLEVIATION

Self-Employment Programmes

Integrated Rural Development Programme (IRDP) & Allied Programmes

The Integrated Rural Development Programme (IRDP) aims at providing self-employment to the rural poor through acquisition of productive assets or appropriate skills which would generate additional income on a sustained basis to enable them to cross the poverty line. Assistance is provided in the form of subsidy and bank credit. The target group consists largely of small and marginal farmers, agricultural labourers and rural artisans living below the poverty line. The pattern of subsidy is 25 per cent for small farmers, 33-1/3 per cent for marginal farmers, agricultural labourers and rural artisans and 50 per cent for Scheduled Castes/Scheduled Tribes families and physically handicapped persons. The ceiling for subsidy is Rs.6000/-for Scheduled Castes/Scheduled Tribes families and the physically handicapped; for others, it is Rs.4000/-in non-DPAP/non-DDP areas and Rs.5000/-in DPAP and DDP areas. Within the target group, there is an assured coverage of 50 per cent for Scheduled Castes/Scheduled Tribes, 40 per cent for women and 3 per cent for the physically handicapped. Priority in assistance is also given to the families belonging to the assignees of ceiling surplus land, Green Card Holders covered under the Family Welfare Programme and freed bonded labourers.

IRDP is a Centrally Sponsored Scheme which is in operation in all the blocks of the country since 1980. Under this scheme Central funds are allocated to States on the basis of proportion of rural poor in a State to the total rural poor in the country. Since the inception of the programme till 1996-97, 50.99 million families have been covered under IRDP at an expenditure of Rs.11434.27 crore. The total investment during this period has been Rs.28047.65 crore which includes a subsidy component of Rs.9669.97 crore and a credit disbursement of Rs.18377.68 crore. Of the total families assisted under this programme 44.75 per cent were Scheduled Castes/Scheduled Tribes and 27.07 per cent women.

During the Eighth Five Year Plan the total allocation (Centre and State) under IRDP was Rs.5048.29 crore and the total investment amounted to Rs.11541.06 crore. In quantitative numbers, 10.82 million families were covered under IRDP against the initial target of 12.6 million families fixed for the entire Eighth Plan period. However, from 1995-96 physical targeting under the programme was abolished with the focus shifting to financial targets and qualitative parameters. Of the families covered 50.06 per cent were Scheduled Castes/Scheduled Tribes and 33.59 per cent women. The coverage of women was still lower than the target of 40 per cent.

The IRDP has been successful in providing incremental income to the poor families, but in most cases the incremental income has not been adequate to enable the beneficiaries to cross the poverty line on a sustained basis mainly because of a low per family investment. The results of the last Concurrent Evaluation (September 1992-August 1993) revealed that of the total beneficiaries assisted under the programme, 15.96 per cent of the old beneficiary families could cross the revised poverty line of Rs.11,000 (at 1991-92 prices), while 54.4 per cent of the families were able to cross the old poverty line of Rs.6,400 per annum. But, the analysis by income group of families revealed that in case of those within initial income of Rs.8501 – 11,000, 48.22% of beneficiary families could cross the poverty line of Rs.11,000 which is quite encouraging. The analysis of the family income of the beneficiaries reveal that a large percentage (57.34%) of the families had annual family income from assets of more than Rs.2000. The annual income from the asset was more than Rs.6000 in 29% cases.

The major constraint in the implementation of IRDP has been sub-critical investments which have adversely affected the Incremental Capital Output Ratio (ICOR) levels and thereby undermined the viability of the projects. Though the average per family investment has been rising steadily in monetary terms, in real terms the increase has been inadequate and in some cases sub-critical due to the inflationary trends and the increase in the cost of assets.

At the instance of the Ministry of Rural Development (now renamed as Ministry of Rural Areas & Employment), the Reserve Bank of India appointed in 1993, a High Powered Committee under the Chairmanship of Dr. D.R. Mehta, Deputy Governor of Reserve Bank of India to make an indepth study of IRDP and recommend suitable measures for its improvement. The Committee was asked to review among other factors, the process of selection of appropriate income generating assets, credit structure, recovery of loans, and procedural matters in respect of obtaining loans, and efficacy of existing administrative structures of the District Rural Development Agencies (DRDAs). In consonance with the recommendations of the High Powered Committee, the new initiatives taken by Government under IRDP in the Eighth Plan included (a) targeting the segment of literate unemployed youth below the poverty line for IRDP activities by giving them subsidy upto Rs.7500 or 50 per cent of the project cost (whichever is lower) (b) promotion of group activities through enhancement of ceiling on subsidy to Rs.1.25 lakh or 50 per cent of the project cost (whichever is lower) for all group ventures involving at least 5 members (c) back-ending of subsidy to prevent leakages in subsidy administration (d) shifting the emphasis to financial targets and qualitative parameters from a perfunctory physical coverage of families and (e) enhancing the limit of allocation to programme infrastructure from 10 per cent to 20 per cent in all the States and 25 per cent in the North Eastern States.

Among the other steps taken to enhance the efficacy of the programme are abolition of the cut of line to enable all families below the poverty line to be assisted under the programme, targeting the investment per family at progressively higher levels each year, extension of the family credit plan to 213 districts of the country, enhancing the ceiling limit of collateral free loans to a uniform limit of Rs.50,000 with a view to easing the constraints faced by poor beneficiaries while taking loans from the banks, extension of the cash disbursement scheme to 50 per cent blocks in the country, decentralisation of the sanctioning powers for infrastructural projects below Rs.25 lakh and entrusting the banks with the task of identification of beneficiaries in about 50 districts on a pilot basis. These interventions have had an impact on the average per family investment which rose from Rs.7889 in 1992-93 to Rs.15036 in 1996-97.

In pursuance of the High Powered Committee's recommendation, for the first time in 1995-96 credit targets were fixed. There has been a continuous increase in the volume of credit mobilised by the banks during the successive years of the Eighth Plan period. Correspondingly, the subsidy credit ratio, which averaged 1:1.77 in the first three years of the Eighth Plan, rose to 1:1.96 in the fourth year and further to 1:2.17 in 1996-97. However, there are genuine reasons for the inability of the banks to meet the full credit requirements of IRDP beneficiaries. These include poor recovery of IRDP loans, lack of adequate rural banking infrastructure in certain areas and the weak financial performance of Regional Rural Banks and Cooperative banks.

There has been considerable diversification of IRDP activities since the inception of the programme. Initially, a majority of the beneficiaries under the programme subscribed to primary sector activities. In 1980-81 the sectoral composition of IRDP activities was heavily skewed towards the primary sector which had a sponsorship of 93.56 per cent, while the share of the secondary and tertiary sectors were 2.32 per cent and 4.12 per cent respectively. Over the years, the share of the primary sector has come down considerably and is currently around 55 per cent, while the shares of the secondary and tertiary sectors have increased proportionately to 15 per cent and 30 per cent respectively.

Inadequate development of infrastructure and insufficient forward and backward linkages and market facilities have been an area of concern under IRDP. In an attempt at filling up the critical infrastructural gaps and strengthening the linkages and marketing facilities, the allocation under IRDP towards the development of programme infrastructure was increased from 10 per cent to 20 per cent in all the States and to 25 per cent in the North Eastern States. Decentralisation in the sanctioning powers for infrastructural projects had already been given effect to in

1994-95. However, despite this enhanced provision for programme infrastructure under IRDP and the relaxation in sanctioning norms, the actual expenditure on infrastructural development was a mere 5 per cent to 7 per cent of the total allocation under the programme at the all-India level. There is, therefore, a critical need to prepare a perspective infrastructural plan at the district and block level and to ensure that the funds earmarked for infrastructural development under IRDP are closely monitored and not diverted elsewhere.

The salient features of IRDP performance during the Eighth Plan are given in Annexure-I.

TRAINING OF RURAL YOUTH FOR SELF-EMPLOYMENT (TRYSEM)

The Scheme of TRYSEM, a facilitating component of IRDP, aims at providing basic technical and entrepreneurial skills to the rural poor in the age group of 18 to 35 years to enable them to take up income generating activities. The Eighth Plan had emphasised the importance of a proper assessment of the training needs of the rural youth in relation to self and wage-employment opportunities, quality of training and group training. During the Eighth Plan, 15.28 lakh youth were trained under TRYSEM, of whom 34.16 per cent took up self-employment and 15.05 per cent wage-employment; while the remaining 50.79 per cent remained unemployed. (Performance details are given at Annexure-II.)

With a view to strengthening this programme, several initiatives were taken in the Eighth Plan which include, among others, an increase in the stipend and honorarium rates; emphasis on professionalised training through the established and recognised institutes like ITIs, Community Polytechnics, Krishi Vigyan Kendras etc., exploring the possibilities of setting up production groups from amongst TRYSEM trainees for undertaking ancillary activities like manufacture and assembly of modern items of production; utilisation of TRYSEM infrastructure funds for the strengthening of Nirmithi Kendras (Rural Building Centres) sponsored by HUDCO for training of youth under TRYSEM in the trades of low cost housing and the setting up of mini-ITIs at the block level to strengthen the training infrastructure for the rural youth.

The TRYSEM programme was evaluated for the first time in a Quick Study (June to August 1993) conducted through independent research institutes/organisations. The main findings of the evaluation study are as under :

i. Of the total sample districts, area skill surveys were not carried out in 92 per cent of the districts to assess the potential skill requirements. This resulted in a mismatch of job skills in 53.3 per cent of the sample districts.

ii. Of the total number of beneficiaries, who got training under TRYSEM, roughly 47.19 per cent were unemployed after the training and 32.54 per cent took up self-employment after training of whom 12.41 per cent took up employment in trades other than those in which they were trained.

iii. A majority of the beneficiaries i.e. 66.52 per cent cited lack of funds as a major reason for not taking up self-employment independently after the training.

iv. A major proportion of TRYSEM trainees i.e. 53.57 per cent did not apply for loan under IRDP. Of the total beneficiaries, who applied for loan, only about 50 per cent were given assistance under IRDP upon completion of training.

v. Roughly, 73.38 per cent of the beneficiaries could derive an average monthly turnover upto Rs.1000 as a result of self-employment taken up by them after the training.

vi. 63 per cent beneficiaries felt no improvement in their socio-economic conditions as a result of TRYSEM training.

There has been a poor convergence of TRYSEM with IRDP which has also been reflected in the Fourth Round of the Concurrent Evaluation of IRDP (1992-93). Only 3.88 per cent of the IRDP beneficiaries had received training under TRYSEM. It was also observed that the rural youth trained under TRYSEM were only interested in the stipendiary benefits they received during the course of training and therefore, had not utilised the knowledge gained under the programme for furthering their self-employment prospects. In practice, therefore, such expenditure on training had become infructuous because of an absence of linkages between the employment opportunities available and training provided. Clearly,

TRYSEM has been a weak link in the overall strategy for self-employment in rural areas.

SUPPLY OF IMPROVED TOOLKITS TO RURAL ARTISANS (SITRA)

Launched in July 1992, as a sub-scheme of IRDP in selected districts, this scheme has since been extended to all the districts of the country. Under the scheme, a variety of crafts persons, except weavers, tailors, needle workers and beedi workers, are supplied with a kit of improved hand tools within a financial ceiling of Rs.2000, of which the artisans have to pay 10 per cent and the remaining 90 per cent is a subsidy from the Government of India. The supply of power driven tools, subject to a ceiling of Rs.4500, is also permitted under this scheme. Beyond this, any additional finance required by the artisans can be provided through loans under IRDP. The rural artisans are trained under TRYSEM for which an age relaxation has been provided to them.

Since the inception of this scheme in 1992-93 upto 1996-97, 6.10 lakh toolkits have been distributed to rural artisans at an expenditure of Rs.116.19 crore. (Performance details are given at Annexure-III.) Reports from the State Governments indicate that the scheme has been well received by rural artisans. The more popular crafts under this scheme are blacksmithy, carpentry, stone craft, leather work, pottery and cane & bamboo work. Prototypes of improved tools in these crafts have been developed by the National Small Industries Corporation (NSIC), Regional Design and Technical Development Centres under the Development Commissioner, Handicrafts and other organisations. The SITRA was evaluated by an independent research organisation, i.e. Development Alternatives, New Delhi, in two Districts of Uttar Pradesh, namely Agra and Aligarh. The findings of this study reaffirms the positive impact of SITRA. It also indicates that the income level of rural artisans have increased substantially with the use of improved tools.

DEVELOPMENT OF WOMEN AND CHILDREN IN RURAL AREAS (DWCRA)

The special scheme for Development of Women and Children in Rural Areas (DWCRA) aims at strengthening the gender component of IRDP. It was started in the year 1982-83, on a pilot basis, in 50 districts and has now been extended to all the districts of the country. The details of the performance under DWCRA during the Eighth Plan are given at Annexure-IV.

DWCRA is directed at improving the living conditions of women and, thereby, of children through the provision of opportunities for self-employment and access to basic social services. The main strategy adopted under this programme is to facilitate access for poor women to employment, skill upgradation, training, credit and other support services so that the DWCRA women as a group can take up income generating activities for supplementing their incomes. It seeks to encourage collective action in the form of group activities which are known to work better and are more sustainable than the individual effort. It encourages the habit of thrift and credit among poor rural women to make them self-reliant. The programme also envisages that this target group would be the focus for convergence of other services like family welfare, health care, nutrition, education, child care, safe drinking water, sanitation and shelter to improve the welfare and quality of life of the family and the community.

Since the inception of the scheme till 1996-97, 1,87,918 DWCRA groups were formed at an expenditure of Rs.248.95 crore, covering 30,39,383 rural women. It was in the Eighth Plan that DWCRA received a fillip with the Government taking several initiatives to strengthen the programme. These include, among others, extending its coverage to all the districts of the country, increasing the revolving fund from Rs.15,000 to Rs.25,000, permitting the formation of smaller DWCRA groups in difficult terrain and remote areas, and permitting operation of joint accounts by the group organiser and another member of the group elected as treasurer of the group rather than the Gram Sevikas and the group organiser, so as to facilitate the DWCRA groups in managing their own affairs. The Child Care Activities (CCA) component was introduced in the DWCRA programme in 1995-96 with the objective of providing child care services for the children of DWCRA women. Similarly the Information, Education and Communication (IEC) component was introduced to generate an awareness among rural women about the development programmes being implemented for their upliftment and welfare. The Eighth Plan also saw the extension of the Community Based Convergent Services (CBCS), a component of DWCRA, to 141 districts of the country.

DWCRA – The Case of Andhra Pradesh

- Strategy adopted for formation of sustainable DWCRA groups had the following salient features :
- Formation of Thrift and Credit groups to develop group dynamics, cohesion and homogeneity among the members.
- Savings provided the entry point for poor women to come together through a Self-Help mechanism.
- Democratically managed groups with collective decision making.
- Sustainable income generating activities with access to credit under the Integrated Rural Development Programme (IRDP) and to training facilities.
- Total Literacy Campaign (TLC), Kalajatha, multimedia publicity campaign through All India Radio (AIR). Doordarshan and print media, involvement of youth leaders, mahila mandals, voluntary organisations and Government functionaries created awareness and contributed to the process of social mobilisation

In the implementation of DWCRA, some States like Andhra Pradesh, Kerala, Tripura and Gujarat have performed very well while in other States, the performance and impact of DWCRA has been relatively poor. In Andhra Pradesh, in particular, several successful DWCRA groups have been formed and this has led to the empowerment of women in decision making on various social aspects that impinge on their daily life. The range of activities pursued by these groups are also fairly diverse. Some have started mini banks and have, thereby, reduced their dependence on the money lenders. Other groups are managing lands taken on lease. Quite a few have formed mini transport companies, having acquired autos, LCVs etc. on bank loans. The success of this programme has been attributed to two major catalysts namely, adult literacy among women and its culmination into a women's movement and close involvement of the NGOs. There is a need to evolve an institutional mechanism for replicating the successful DWCRA groups throughout the country.

PODUPULAKSHMI – Pride of Nellore Women'-a success story.

- As a sequel to the Total Literacy Campaign & anti-arrack agitation, 'PODUPULAKSHMI' (Podupu means saving, Lakshmi – Goddess of Wealth) was started by two lakh women organised into 7000 small thrift groups of 20-30 members. They saved Rs.8.00 crore in four years.
- The district administration provided the 'PODUPULAKSHMI' movement with the initial support and acted as a facilitator. Volunteers, animators, trainers and instructors in the Total Literacy Campaign acted as PODUPULAKSHMI organisers.
- Once PODUPULAKSHMI groups reached a level of maturity, they converted themselves into DWCRA groups. This enabled them to access the revolving fund under DWCRA which was used by the groups to provide working capital to set up micro enterprises. The savings fund was used to meet emergency consumption needs.
- Today a wide variety of women centred activities are carried out by these groups. The ANM, the School Teacher, the Fair Price Shop Dealer, the Anganwadi Worker are all associated with PODUPULAKSHMI – bringing about a convergence of basic services

Yet, in the implementation of DWCRA several shortcomings have also surfaced which has stymied its successful and effective execution in some States. Several groups have become defunct over time. The reasons for these include, among others, (a) improper selection of groups; (b) lack of homogeneity among the group members; (c) selection of non-viable economic activities which are mostly traditional and yield low income; (d) the linkages for supply of raw material and marketing of production are either deficient or not properly planned as a result of which DWCRA groups have become vulnerable to competition. The District Supply and Marketing Societies have been weak outlets for the sale of DWCRA products; (e) lack of institutional financial support, inadequate training, a non-professional approach and poor access to upgraded technological inputs have deprived DWCRA groups from diversifying into high value addition activities; and (f) inadequacy of staff and their insufficient training and motivation has also affected the overall implementation of the programme. These shortcomings would have to be suitably addressed for the successful implementation of the programme in the Ninth Plan.

WAGE EMPLOYMENT PROGRAMMES

Jawahar Rozgar Yojana (JRY)

Rural poverty is inextricably linked with low productivity and unemployment. Hence, it is imperative to improve productivity and increase employment in rural areas. An employment-oriented growth strategy would achieve this goal only in the medium and long run. In the short run supplementary employment will have to be provided to the unemployed and underemployed, during the lean agricultural season. There are two major wage employment programmes namely the Jawahar Rozgar Yojana (JRY) and the Employment Assurance Scheme (EAS) presently in operation.

The JRY was launched as a Centrally Sponsored Schemes (CSS) on 1st April, 1989 by merging the National Rural Employment Programme (NREP) and the Rural Landless Employment Guarantee Programme (RLEGP). The main objective of the programme is the generation of additional gainful employment for unemployed and underemployed persons, both men and women, in the rural areas through the creation of rural economic infrastructure, community and social assets with the aim of improving the quality of life of the rural poor.

The resources under this scheme are allocated to the States/ UTs on the basis of proportion of rural poor in the States/UTs to the total rural poor in the country. From the States to the districts, the allocation is made on an index of backwardness which is based on the proportion of rural Scheduled Castes/Scheduled Tribes population in the district to total Scheduled Castes/Scheduled Tribes population in the State and an inverse of agricultural production per agricultural worker in that district, in equal weights. The funds are devolved to the village panchayats by giving due weightage to the Scheduled Castes/Scheduled Tribes population and the total population of the village panchayat. Until recently, these funds were distributed between the village panchayats and the district level in the ratio of 80:20. However, subsequent to the revitalisation of PRIs at three levels, the JRY funds are now distributed among the village panchayats, intermediate panchayats and the district panchayats in the ratio of 70:15:15.

This programme is targeted at people living below the poverty line. However, preference is given to Scheduled Castes/Scheduled Tribes and freed bonded labourers. Atleast 30 per cent of the employment is to be provided to women under the Yojana. In practice, however, this programme is self targeting. Given that employment is offered at statutory minimum wages for unskilled labour and that these wage rates are generally lower than the prevailing market wage rates, only those willing to do manual work for the prescribed wage rates would seek employment on these public works. While works under the scheme can be taken up during any part of the year whenever the need for generating employment is felt, these should preferably be started during the lean agricultural season but may continue thereafter, if necessary.

After three years of its implementation, i.e. in 1992-93, a review of the programme revealed that the per person employment generated was inadequate in terms of the requirement and did not provide enough income to the poor. It was also perceived that the resources under JRY were too thinly spread and adequate attention was not being given to the backward areas of the country. Accordingly, the strategy for implementation of JRY was modified from 1993-94 with the introduction of the Second Stream of JRY, specifically targeted at 120 identified backward districts in 12 States of the country, characterised by a concentration of the poor and the underemployed, with additional resources flowing to these districts. This modification in programme strategy was made to achieve the target of providing 90-100 days of employment per person in backward districts where there was a concentration of unemployed and underemployed persons. In addition, a Third Stream of JRY was introduced for taking up special and innovative projects aimed at preventing migration of labour, enhancing women's employment and undertaking special programmes through voluntary organisations for drought proofing etc.

A Concurrent Evaluation of the JRY was conducted from June 1993 to May 1994. The study revealed that nearly 82.16 per cent of the available funds were spent on community development projects. Construction of rural link roads received the highest priority. The wage and non-wage component of the expenditure of JRY works undertaken by the village panchayats was of the order of 53:47 at the all-India level against the stipulated norm of 60:40. Muster rolls were maintained with 86.87 per cent of the village panchayats. The average wages paid per manday of the unskilled workers were more or less on the lines of the minimum wages stipulated under the Act. Of the assets created, 76.96 per cent were created by the village panchayats and 76.11 per cent of these assets were found to be in a good condition. As many as 69.35 per cent of the workers were satisfied with the benefits they received under the JRY.

The Evaluation Report also brought into focus certain inadequacies in the programme. It was reported that 57.44 per cent of the elected panchayat heads had not been imparted any training for the implementation of JRY works. The share of women in employment generated under the programme was only 16.59 per cent and 49.47 per cent of the works could not be completed on time on account of shortage of funds. Other shortcomings observed were differentials in the wages paid to male and female workers, non-utilisation of locally available material in a large number of JRY works undertaken by panchayats and lack of discussion of the annual action plans in the Gram Sabha meetings etc.

In a comprehensive restructuring of the wage employment programmes on 1.1.1996, JRY was further streamlined. In the revised strategy, the First Stream of JRY was continued in its existing form but Indira Awaas Yojana (IAY) and Million Wells Scheme (MWS) which were till then sub-schemes of JRY were made independent schemes. The Second Stream of JRY, which was being implemented in 120 backward districts in the country, was merged with the Employment Assurance Scheme (EAS) introduced in 1775 selected backward blocks of the country in 1993-94 in view of the similarity in these programmes. The Third Stream of JRY with its thrust on innovative projects was continued. Accordingly, the JRY is now being implemented in two parts i.e. (i) the Jawahar Rozgar Yojana

(Main Scheme); and (ii) Special and Innovative Project. Since the inception of JRY in 1989-90 till 1996-97 a total amount of Rs.26570.25 crore (Centre and State) was allocated to the programme. As against this total allocation, an amount of Rs.25661.70 crore was released of which Rs.25190.30 crore was utilised by the States. This utilisation is approximately 98.16 per cent of the total funds released. In terms of physical performance, as against a target of 6581 million mandays fixed under JRY, the actual employment generated was 6585 million mandays which is an achievement of 100.07 per cent. Of the total employment generated under the programme, the share of Scheduled Castes/Scheduled Tribes was 3659.53 million mandays (55.57 per cent) and that of women 1681.40 million mandays, which is 25.53 per cent.

The financial and physical performance under the JRY during the Eighth Plan period is given at Annexure-V.

From the data on mandays of employment generated under JRY it is difficult to assess the number of workers who have actually received employment in the rural areas and on an average for how many days. To surmount this shortcoming, registration of workers who take up employment under this programme, should be made compulsory as under the EAS.

Besides generating supplementary employment of a casual manual nature, the programme has contributed to the development of rural infrastructure through the creation of a wide range of community and social assets in a number of sectors. These included major irrigation works, soil conservation works, land development, drinking water wells, rural roads, construction of school buildings, panchayat ghars, mahila mandals, houses and sanitary latrines and social forestry. In fact, assistance for construction of class rooms under the Operation Black Board (OBB) programme was specially provided under JRY. Of the total cost, 60 per cent was funded from JRY and 40 per cent from the education department of the State Governments to meet the additional non-wage cost. Under Operation Black Board, 25576 classrooms and 21541 school buildings were constructed at an expenditure of Rs.176.11 crore from JRY funds between 1991-92 to 1994-95. Such an integration between sectoral programmes and JRY, with dovetailing of funds, would help in the creation of better quality durable assets.

Employment Assurance Scheme (EAS)

The Employment Assurance Scheme was launched on 2nd October, 1993 in 1775 identified backward blocks situated in drought prone, desert, tribal and hill areas, in which the revamped public distribution system was in operation. Subsequently, the scheme was extended to additional blocks which included the newly identified Drought Prone Area Programme (DPAP)/Desert Development Programme (DDP) blocks, Modified Area Development Approach (MADA) blocks having a larger concentration of tribals, and blocks in flood prone areas of Uttar Pradesh, Bihar, Assam and Jammu & Kashmir. In addition, 722 non-EAS blocks previously covered under Second Stream of Jawahar Rozgar Yojana (JRY) were also brought under the EAS. The EAS has since been universalised to cover all the rural blocks in the country with effect from 1.4.1997.

The main objective of the EAS is to provide about 100 days of assured casual manual employment during the lean agricultural season, at statutory minimum wages, to all persons above the age of 18 years and below 60 years who need and seek employment on economically productive and labour intensive social and community works. The works are to be selected by the District Collector and implemented through the line departments in such a manner that the ratio of wage to the non-wage component would stand at 60:40. Sectoral norms for execution of various works are- watershed development (50 per cent) and agro-horticulture, minor irrigation works (10 per cent) in DPAP and DDP blocks or water & soil conservation including afforestation, agro-horticulture and silvipasture (40 per cent), and minor irrigation works (20 per cent) in non-DPAP/non-DDP blocks. In addition, funds are also earmarked for link roads featuring in the Master Plans developed in the respective districts for this purpose (20 per cent) and public community buildings in rural areas as per the felt needs of the districts (20 per cent). The village panchayats are involved in the registration of persons seeking employment and the panchayats maintain these registers. They also coordinate and monitor the works. A maximum of two adults per family are to be provided employment under the scheme. The applicants, who register themselves for employment under the EAS, are issued family cards in which the number of days of employment are entered as and when such employment is given to them.

The EAS is a Centrally Sponsored Scheme. The scheme is demand-driven and therefore no fixed allocations are made for the districts/blocks. Instead, initial notional allocations are made to districts at the commencement of each year and thereafter depending on the demand for supplementary employment and the actual utilisation of funds the districts can request for additional funds. For the purpose of initial release, the blocks have been classified intc three categories i.e. category A, B & C and Central funds to the tune of Rs.40 lakh, Rs.30 lakh and Rs.20 lakh are released as the first of the two instalments to these blocks respectively. This corresponds to the notional minimum allocation of Rs.1 crore, Rs.75 lakh and Rs.50 lakh per block per annum including the State's matching share.

The financial and physical performance of EAS since its inception in 1993-94 (October 1993) to 1996-97 is given in the Annexure-VI.

Since the inception of EAS in 1993-94 (i.e. October 1993) upto 1996-97, a total amount (Centre and State) of Rs.6514.65 crore has been released under the programme, against which the total utilisation was Rs.5278.16 crore. This indicates a percentage utilisation of 81.02 per cent. As many as 25.90 million persons registered themselves for employment under the EAS. The programme generated 1068.60 million mandays of employment from 1993-94 (October 1993) to 1996-97.

The EAS has not been evaluated till date. Hence, it is difficult to assess the overall impact of the programme in terms of employment, enhancement in the purchasing power of the poor and creation of durable assets. However, the Programme Evaluation Organisation has recently undertaken a comprehensive evaluation study of this Scheme.

Million Wells Scheme (MWS)

In India, though the small and marginal farmers, with holdings of less than 2 hectares, account for about 78 per cent of the total operational holdings, they only cultivate about 32.2 per cent of the cropped area (Agricultural Census 1990-91). To increase the productivity of these holdings they must be ensured an assured source of water supply. Ground water made available through wells is an important source specially in the remote areas of the countryside, where canal or tank irrigation is not feasible. Though the fixed capital investment in well irrigation is fairly high, it has many advantages such as flexibility in operation, dependability of source, timing of water deliveries and low conveyance losses.

The Million Wells Scheme (MWS) was launched as a sub-scheme of the National Rural Employment Programme (NREP) and the Rural Landless Employment Guarantee Programme (RLEGP) during the year 1988-89. After the merger of the two programmes in April 1989 into the Jawahar Rozgar Yojana (JRY), the MWS continued as a sub-scheme of JRY till December 1995. The MWS was delinked from JRY and made into an independent scheme with effect from 1.1.1996.

The scheme was primarily intended to provide open irrigation wells, free of cost, to individual, poor, small and marginal farmers belonging to Scheduled Castes/Scheduled Tribes and freed bonded labourers with a 20 per cent earmarking of JRY funds. Tubewells and borewells are not permitted under the Scheme. Where wells are not feasible due to geological factors, other minor irrigation works can be undertaken such as irrigation tanks, water harvesting structures as also development of land belonging to small and marginal farmers. From the year 1993-94 the scope of the MWS has been enlarged to cover non-Scheduled Castes/non-Scheduled Tribes small and marginal farmers who are below the poverty line and are listed in the IRDP register of the village. The sectoral earmarking which was 20 per cent upto 1992-93 had also been raised to 30 per cent from 1993-94 with the stipulation that the benefits to non-Scheduled Castes/Scheduled Tribes would not exceed one third of the total funds utilised during the year.

The MWS is also a Centrally Sponsored Scheme. The cost/area norms in regard to works under MWS are decided upon by a Committee comprising of Chief Secretary, Secretary (RD), Secretary (Planning), Secretary (Irrigation) and Chief Engineer (Minor Irrigation) of the State. The beneficiarficiaries themselves are asked to undertake construction of their wells through their own labour and local labour for which they are paid. Contractors are banned under this programme. The wage to material ratio is required to be maintained at 60:40. Supplementary material costs, if any, can be met from other private/public sources. Though lifting devices are not provided under the scheme, the beneficiaries who intend to install a lifting device, are given the preference under IRDP and other relevant programmes.

The MWS is being implemented throughout the country. Allocations are made to the States/UTs on the basis of the proportion of rural poor in the State/UTs to the total rural poor in the country. The District-wise allocations are made by the States from their allocation in relation to the unirrigated land held by the target group with a potential for well irrigation.

A total of 11.04 lakh wells have been constructed since the inception of the programme till 1996-97 at an expenditure of Rs.4003.11 crore. The financial and physical performance under MWS during the Eighth Plan is given at Annexure-VII.

There has been no evaluation or impact study conducted in the field for the MWS. Yet, on the basis of the feedback available from certain parts of the country, this programme achieved considerable success in the districts falling in the Chotanagpur region of South Bihar, large parts of Orissa, many districts of Gujarat, besides the Eastern and Southern region of Rajasthan. In these areas, the MWS has played a significant role in transforming single cropped dry land areas held by farmers of the target group into double cropped lands, leading to increase in agricultural output and incomes. Yet, such successes have not been uniformly reported across the country. Many States have expressed difficulty in the implementation of the programme. For instance, in Punjab and Haryana where the incidence of tubewell irrigation is widespread and there is a wide network of canal irrigation systems, the programme of open dug wells is a non-viable option. Similarly, in Kerala the small size of the land holdings of the small and marginal farmers gives the scheme a limited potential. Andhra Pradesh, Goa, Madhya Pradesh, Maharashtra, and West Bengal are some of the other States which have shown a poor performance in the construction of wells under the MWS. These States have been permitted to utilise the allocations made under MWS for other schemes of minor irrigation such as irrigation tanks, water harvesting structures and also for the development of land belonging to the small and marginal farmers. Some States have also been permitted to divert MWS resources for the construction of houses for the poor under the Indira Awaas Yojana.

Field studies in various parts of the country have identified several factors which have posed as impediments to the effective implementation of this scheme. These include, among others, (a) construction of wells without proper hydro-geological surveys; (b) a declining water table and its continuous depletion by overuse of pumping sets resulting in large tracts falling in the dark/grey zones which indicate already dangerous levels of depletion of ground water; (c) non-availability of eligible persons in the target group; (d) limited success in rocky and sandy strata; and (e) distance between wells affecting the rate of discharge.

The programme has also been hamstrung by inadequate linkages. There has been a failure on the part of the block officers and the banks in providing lifting devices under IRDP and other programmes, thus rendering the investment in open dug wells infructuous. In some cases, though the wells have been dug and are working, their full potential has not been realised because of a lack of extension support from the agricultural department.

National Social Assistance Programme (NSAP)

The National Social Assistance Programme (NSAP) came into effect from 15th August, 1995. The programme represents a significant step towards the fulfilment of the Directive Principles in Articles 41 and 42 of the Constitution through the enunciation of a National Policy for social assistance benefits to poor households in the case of old age, death of the primary breadwinner and maternity. It is a Centrally Sponsored Scheme with 100 per cent Central assistance provided to States/UTs.

This programme has three components : namely (i) National Old Age Pension Scheme (NOAPS); (ii) National Family Benefit Scheme (NFBS); and (iii) National Maternity Benefit Scheme (NMBS) which are targeted at people living below the poverty line. Under the National Old Age Pension Scheme (NOAPS), old age pension of Rs.75 per month is provided to persons of 65 years and above who are destitutes. The National Family Benefit Scheme

(NFBS) provides a lump sum family benefit of Rs.10,000 to the bereaved household in case of the death of the primary bread winner irrespective of the cause of death. This scheme is applicable to all the eligible persons in the age group 18 to 64. Under the National Maternity Benefit Scheme (NMBS) there is a provision for payment of Rs.500 per pregnancy to women belonging to poor households for pre-natal and post-natal maternity care upto the first two live births. This benefit is provided to eligible women of 19 years and above.

In providing social assistance benefits to poor households in cases of old age, death of the primary bread winner and maternity, the NSAP supplements the efforts of the State Governments with the objective of ensuring minimum national levels of well-being.

The NSAP provides opportunities for linking social assistance package to schemes for poverty alleviation and provision of basic minimum services. In fact, that old age pension can be linked to medical care and other benefits aimed at the aged beneficiaries. The Integrated Rural Development Programme/Jawahar Rozgar Yojana assistance may be provided in addition to the family benefit for the families of poor households, who suffer the loss of the primary bread winner. Maternity assistance can be linked to other programmes of maternal and child care.

The NSAP is implemented in the States/UTs through Panchayats and Municipalities. The Panchayats and Municipalities are encouraged to involve Voluntary agencies to the extent possible for identifying beneficiaries and persuading them to avail of the benefits intended for them.

The NSAP programme has been operationalised since August 1995. No evaluation of the programme has been conducted either by the Central Government or by any other institution/organisation so far.

The details of financial and physical performance under NSAP during the years 1995-96 and 1996-97 are given at Annexure VIII.

In the first year of the programme in 1995-96, as against the total allocation of Rs.538.93 crore, the release was Rs.380.49 crore. Of the total release, only Rs.183.77 crore was actually utilised. Consequently, in 1996-97 there was a large opening balance of

Rs.197.53 crore. Again, as against the total allocation of Rs.916.21 crore, the total release in 1996-97 was Rs.548.29 crore. Thus, the total available funds in 1996-97 was Rs.745.82 crore, of which only Rs.383.50 crore was utilised (i.e. 51.42 per cent utilisation).

Of these three programmes, the performance under the NOAPS has been relatively good as compared to that of NFBS and NMBS because the administrative machinery for the implementation of this programme was already in place in most of the States. If we analyse the State-wise physical performance for the year 1996-97, even under NOAPS, only 10 States had covered more than 90 per cent numerical ceilings of the target population. Of this, 5 States/UT namely Andhra Pradesh, Haryana, Karnataka, Madhya Pradesh and Pondicherry had more than 100 per cent achievement. The achievement in many States was less than 50 per cent. These States include Goa, Maharashtra, Rajasthan and North Eastern States excluding Assam.

The performance in the case of NFBS and NMBS was particularly poor. Under NFBS, achievement was only 28.85 per cent of the numerical ceiling. In only five States namely, Andhra Pradesh, Goa, Madhya Pradesh, Punjab and Tamil Nadu the coverage was above 50 per cent and in 7 States/UTs the coverage was between the national average of 28.85 per cent and 50 per cent, while in the remaining 20 States/UTs, the coverage was below the national average.

Under NMBS, achievement was only 27.57 per cent of the numerical ceiling. Except in Andhra Pradesh and Tamil Nadu where the coverage was 51.36 per cent and 60.94 per cent respectively, in all other States/UTs the coverage was below 50 per cent. In the case of 7 States/UTs the coverage was between the national average of 27.57 per cent and 50 per cent, while in the remaining 23 States/UTs, the coverage was below the national average.

Steps would be taken to simplify the procedures for the sanctioning and disbursement of benefits under these schemes with a closer involvement of the Gram Panchayats/Municipalities. The assistance may be sanctioned and disbursed in public meetings preferably of Gram Sabha by either Gram Panchayat functionaries or Block functionaries or appropriate level. In the case of urban beneficiaries, elected local self government officials wherever available should be involved in the process of sanctioning the assistance. The disbursement in such cases should also be preferably made in public meetings of neighbourhood/mohalla committees. These schemes would also be given increased publicity to ensure a larger coverage of beneficiaries.

LAND REFORMS

Despite attempts at land reforms over successive Plan periods, the basic character of the agrarian economy has not undergone any structural change. The pattern of land distribution is highly skewed, with a high concentration of land in the hands of a few land owners on the one hand and the growing number of marginal and sub-marginal farmers on the other. Fragmentation of land holdings continues on a large scale and only a few States like Punjab, Haryana, Uttar Pradesh and parts of Maharashtra have been able to successfully undertake a programme of consolidation of holdings.

Agricultural tenancy, which was abolished in most of the States by various enactments in the post-Independence era, continues unabated though it is largely concealed. In the wake of liberalisation, several State Governments have modified their land ceiling laws so as to exempt orchards, fish ponds etc., from the purview of land ceilings. There is also a move to make suitable changes in tenancy regulations to attract private corporate investment in agriculture. Hence, it is necessary to reconsider the issue of land reforms, particularly from the point of view of the poor, as access to land is still a major source of livelihood in rural India. In fact, it has been argued that the need for poverty alleviation programmes has arisen because the land reforms have not been implemented in a systematic way. The experience of several countries in East Asia shows that land reforms, leading to structural equity in the distribution of land, are an essential prerequisite for economic development through agricultural transformation. In addition, the efficiency of land use and land management, and protection of the land rights of the tribals and women have assumed great significance in the context of the changes that are taking place in rural India.

The continued importance of land reforms was recognised in the Eighth Plan, with the abolition of intermediaries, redistribution of ceiling surplus land, tenancy reforms providing security of tenure to tenants and share croppers, consolidation of holdings and updating of land records as the main objectives of the land reform policy. However, only limited success was achieved with respect to these objectives in the Eighth Plan. Given that land reforms is a State subject, the Central Government can only draw the attention of the State Governments to the pressing needs for land reforms, which are central to any strategy of poverty alleviation.

At the end of the Seventh Plan, out of the 72.2 lakh acres of land declared surplus, only 46.5 lakh acres had been distributed. At the end of the Eighth Plan, out of the total 74.94 lakh acres declared surplus, 52.13 lakh acres had been distributed. In other words, during the Eighth Plan only 6-7 lakh acres were redistributed. Further, 12.4 lakh acres were under disputes pending in courts and 19.59 lakh acres were not available for distribution because they were unfit for cultivation or reserved for public purposes or for other miscellaneous reasons. In fact, only 59,000 acres were available for redistribution. Of the Bhoodan land donated, 53 per cent was distributed, accounting for 24.52 lakh acres. In addition, 142.87 lakh acres of wastelands were distributed among 88.5 lakh beneficiaries. But, there is still considerable scope for redistributing Government wastelands, common lands, ceiling surplus land and Bhoodan land.

Similarly, in the area of tenancy reforms very little progress has been made, after the initial abolition of 'zamindari' and the transfer of title to owner-cultivators in the immediate post-Independence period. The successful implementation of tenancy laws has been confined to West Bengal, Karnataka and Kerala. In fact, in the Eighth Plan there was no progress in respect of conferment of rights on tenants and therefore the issue of tenancy reforms is still illusory, but requires tackling.

Consolidation of holdings has taken place in very few States. While 15 States had enacted appropriate legislation, Andhra Pradesh, Tamil Nadu, Kerala, Pondicherry and the North-Eastern States do not have any laws for consolidations of holdings. Several

States like Bihar, Maharashtra and Rajasthan have suspended the programme. In fact, only in Uttar Pradesh, 900-1000 villages are being covered annually.

There is evidence of considerable alienation of tribals from their land. As per the latest available estimates, 4.6 lakh cases of tribal land alienation covering 9.2 lakh acres have been registered. Of these 2.7 lakh cases covering 6.3 lakh acres have been disposed of in favour of tribals but physically an estimated 4.7 lakh acres had been restored to them. In other cases, reconciliations are being effected.

It cannot be gainsaid that the essential prerequisite of any land reform measure is the recording of land rights and their updating. In recognition of this, a Centrally Sponsored Scheme for Strengthening of Revenue Administration and Updating of Land Records was introduced during the later half of the Seventh Plan and against an outlay of Rs.20.8 crore, Rs.14 crore were spent. In the Eighth Plan a provision of Rs.175 crore was made against which Rs.98 crore were released and Rs.66.7 crore utilised.

But, given that the funds are shared in the ratio of 50:50 between Centre and States, several States have not been able to provide their share and hence, the utilisation has been low. Several other States have not availed of this scheme at all. In 1995-96, funds were released only to Bihar, Kerala, Maharashtra, Madhya Pradesh, Rajasthan, Tamil Nadu and West Bengal aggregating to only Rs.18.8 crore.

The States have been requested repeatedly to expedite expenditure and to adopt new technologies for survey and settlement operations, preparation of maps etc. Funds for infrastructure development were also sanctioned to meet the training needs of the revenue functionaries. Some States have developed new training institutions, while a few have upgraded the existing ones.

In 1988-89, 8 pilot projects were taken up for computerisation of land records with 100 per cent Central assistance. Since inception, 323 projects have been sanctioned and funds to the tune of Rs.64.43 crore have been released. However, the utilisation was only to the extent of 20 per cent. This shows the tardy progress made in the implementation of this scheme on the ground.

NINTH PLAN STRATEGY

Direct poverty alleviation programmes will continue on an expanded scale in the Ninth Plan. But these programmes would be oriented towards strengthening the productive potential of the economy and providing more opportunities for involving the poor in the economic process.

The Integrated Rural Development Programme (IRDP) would continue to be the major self-employment programme targeted to families living below the poverty line in the rural areas of the country. In the Ninth Plan it would be implemented through an integrated approach under which the existing schemes of Training of Rural Youth for Self-Employment (TRYSEM) and Supply of Improved Toolkits to Rural Artisans (SITRA), Development of Women and Children in Rural Areas (DWCRA) and Ganga Kalyan Yojana (GKY) would be subsumed into the main programme.

To facilitate higher levels of investment under the programme, there would be a strategic shift under IRDP from an individual beneficiary approach to a group and/or cluster approach. As part of the group approach, the focus of IRDP would be on the formation of Self-Help Groups (SHGs) which would be the catalyst for organising the poor. The cluster approach would focus on the identification of a few specified viable activities based on the local resource endowment and occupational skills of the people of that area. The IRDP will also aim at diversifying the investments into high-value-addition sectors and non-traditional activities which have a market potential. The financial institutions would play a more significant and dynamic role by enhancing the credit flows through a continuous line of credit, instead of a one-time loan and would render constructive assistance to the beneficiaries. This would help them to increase returns on their investments. The IRDP would service the beneficiaries through a package approach, wherein the beneficiary would have access to credit, training as per requirements, upgradation of technology, delivery of essential inputs and marketing tie-ups in an integrated manner. Presently, under the IRDP, training under TRYSEM is provided as an isolated input and there has been little attempt to make a proper assessment of opportunities where skills could be more gainfully utilised. Hence, the new holistic approach would overcome some of the inherent shortcomings which have undermined the success of the programme.

In view of the near complementarity in the ongoing wage employment schemes, Jawahar Rozgar Yojana (JRY) and Employment Assurance Scheme (EAS) will be rationalised. In the revised format JRY will be confined to the creation of rural infrastructure at the village panchayat level in consonance with the felt needs of the community. To the extent that the works undertaken under JRY would be largely labour intensive, supplementary wage employment would be generated in the process of infrastructure creation. The EAS would be the major wage employment programme which would contribute significantly to the provision of the mandated 100 days of casual manual work to those who register for employment under it. As the EAS has been universalised, specific measures would be taken to ensure that the benefits reach the poor and more backward districts of the country.

Land reforms will continue to be an important policy instrument for alleviating rural poverty. Access to land is still a major source of liveliood and its possession enhances the status of people in rural society. A proper implementation of land laws and policies would lead to a restructuring of the agrarian economy in a way conducive to higher rates of agricultural growth but with greater equity in the distribution of gains from it. While the ingredients of the land reform policy would continue to be the same as before, the focus would shift to a few critical areas. All efforts would be made to detect and redistribute the ceiling surplus land and to enforce the ceiling laws stringently. Given that small and marginal farms are viable, both from the efficiency and equity points of view, it is desirable that the existing ceiling limits are strictly enforced. More importantly, tenancy reforms would have to be taken up especially in States characterised by semi-feudal modes of production. The rights of tenants and sharecroppers need to be recorded and security of tenure provided to them. Leasing in of land should be made permissible within the ceiling limits. The poor should be given access to wastelands and common property resources. The land rights of women must be ensured. This would require amendment of the existing legislations in some States to ensure women's rights with regard to inheritance of both land owned as also under tenancy. Updating of land records would have to be expedited as this is a necessary prerequisite of any effective land reform policy. Since land reforms is a State subject the States would have to be persuaded to take up these measures.

While the implementation of poverty alleviation programmes would be made more effective in the Ninth Plan, the extant non-monetary policies and institutional arrangements which adversely affect the interests of the poor would also be suitably addressed. In this context, the Ninth Plan will identify those laws, policies and procedures which are anti-poor and take the initiative in exploring whether some of these could be suitably modified in favour of the poor.

In the Ninth Plan, the poverty alleviation programmes would be more effectively integrated with area development programmes and the various sectoral programmes within the umbrella of Panchayati Raj Institutions (PRIs). The PRIs will function as effective institutions of local self-government. These would prepare the plans for economic development and social justice through the District Planning Committees and implement them. The State Government would have to devolve administrative and financial power to the PRIs which fall within their purview as per the Eleventh Schedule of the 73rd Constitutional Amendment Act, 1992. In addition, the voluntary organisations (VO) will have to play a more dynamic role in empowering the poor through advocacy, awareness generation and formation of SHGs.

Integrated Rural Development Programme (IRDP)

New Initiatives under IRDP

- IRDP will be a holistic programme covering all aspects of self-employment, namely, organisation of beneficiaries and their capacity building, planning of activity clusters, infrastructure, technology, credit and marketing.
- The existing sub-schemes of TRYSEM, DWCRA, SITRA and GKY to be merged into IRDP.
- Progressive shift from the individual beneficiary approach to the group and/or cluster approach.

- To facilitate group approach SHGs will be formed and steps will be taken to nurture them.
- For cluster approach each district will identify 4 to 5 activity clusters in each block based on local resources and occupational skills of the people. The infrastructure needs for the identified activities will be met in full.
- The Banks will be closely involved in the planning and preparation of projects, identification of activity clusters, infrastructure planning as well as capacity building and choice of activity of the SHGs.
- Promotion of multiple credit rather than one time credit injection.

The IRDP would continue to be the major self-employment programme, targeted towards families living below the poverty line in the rural areas. However, in the Ninth Plan, the focus would be on pursuing an integrated approach under IRDP by subsuming the existing sub-schemes of Training of Rural Youth for Self-Employment (TRYSEM) and Supply of Improved Toolkits to Rural Artisans(SITRA), Development of Women and Children in Rural Areas (DWCRA) and Ganga Kalyan Yojana (GKY) into the main programme. This integration of schemes is necessary to develop the appropriate forward and backward linkages to achieve a synergistic complementarity in the overall implementation of the programme.

Project for Linking Banks with Self-Help Groups

- NABARD initiated a Pilot project to link Banks with Self-Help Groups (SHGs), with the objective of meeting the credit needs of the poor through formal financial institutions.
- Evaluation studies show success with increase in loan volume of SHGs, shift to income generating activities, nearly 100 per cent recovery and reduction in transaction costs.
- 85% of the groups linked with banks were exclusively of women.
- Scheme is being expanded, and will be replicated all over the country.

Furthermore there would be a strategic shift from an individual beneficiary approach to a group and /or cluster approach.

a. To facilitate this process Self Help Groups (SHGs) will be formed under IRDP and steps will be taken to nurture these groups to enable them to function effectively as well asto choose their economic activity. Efforts would be made to involve women members in each SHG. Besides, formation of exclusive women groups will also continue as at present under DWCRA. It is proposed that group ventures which involve at least five beneficiaries would be assisted by making available 50 per cent of the project cost subject to a ceiling of Rs.1.25 lakh. This enhanced level of investment would facilitate economies of scale and improve recovery.

b. Alternatively, a cluster approach would be preferred wherein a few specified activities are identified for assistance in an area. This would necessitate the formulation of a menu of "activity-based" project profiles in different sectors to suit the local resource endowment and the occupational skills of the local people. Accordingly, each DRDA would set up four to five activity clusters. Appropriate infrastructure and technology inputs would be built into the project.

c. The Family Credit Plan would be extended to all the districts of the country in a phased manner. The feedback from the States suggests that this strategy has met with reasonable success and has raised investment levels.

One of the major constraints in the implementation of IRDP has been sub-critical investments, which have adversely affected Incremental Capital Output Ratios (ICOR) and, thereby, undermined the viability of the projects. Recognising that the level of investment is the most crucial variable in determining the incremental income generated under IRDP, the credit flows and the average level of investment per family for the Ninth Plan would aim at achieving enhanced levels of investment in the range of Rs.25,000-Rs.50,000 depending on the estimate of the poverty line and the poverty gap. These higher levels of investment will give the beneficiary the necessary financial support for diversifying into high-value-addition sectors and non-traditional activities which have a market potential.

In this effort at achieving higher investment levels, the financial institutions would have to play a more significant and dynamic role by enhancing credit flows and rendering constructive assistance to the beneficiary making their investments viable under this programme. Adoption of simplified procedures by financial institutions would facilitate the BPL families in accessing groups loans under IRDP.

To simplify procedures the regimen of subsidy administration would be suitably streamlined in the Ninth Plan. At present, there is a complex gamut of ceiling limits on subsidy. The ceiling limit on subsidy would be fixed at 30% of the project cost subject to a maximum of Rs.7,500. However, in the case of SC/ST it would be fixed at 50% of the project cost subject to a ceiling of Rs.10,000. Therefore, for purposes of administrative expediency the area specific differential in subsidy administration would no longer exist. For group activities, however, the subsidy would continue to be Rs.1.25 lakh or 50 per cent of the project cost, whichever is lower.

The IRDP will seek to develop close linkages with the credit mechanism in such a manner as would promote repeated/multiple doses of credit rather than a one-time loan for the beneficiary. The emphasis, therefore, would be on establishing a continuous line of credit for the beneficiary, wherein it would be possible for the borrower to obtain need-based additional credit for working capital purposes, meeting unforseen expenditure related to proper maintenance of assets etc. which would have a bearing on the viability of the project so as to sustain credible levels of income generation.

Experience has shown that the IRDP has been relatively more successful in land-based activities. In recognition of this fact, purchase of land was made a permissible activity under the programme. For land-based activities, besides providing assistance for purchase of inputs to enhance the productivity of land, there exists a potential for diversifying into other allied activities, which have a high value-addition, such as sericulture, aqua-culture, horticulture and floriculture, on the existing lands of the small and marginal farmers as well as on land leased by the landless. In addition to these activities, SHGs would be given 'pattas' for development of wastelands, social forestry, soil conservation and watershed projects. Common property resources would also be allocated to SHGs in the villages on a long-term basis for eco-sensitive resource management. There is also a potential for allotting nurseries of the forest department to these groups for management. In all these, interest of women's groups would have to be protected. Usufruct rights of women on minor forest produce would have to be ensured legally and administratively and assistance provided to them for purchase or leasing in of land for joint management.

Yet, it must be recognised that land is a limited resource and that the distribution of land holdings remains highly skewed. Hence, the need for exploring the potential of the non-farm sector is crucial. It cannot be gainsaid that the non-farm sector in the rural areas has witnessed both growth and diversification in the past few years. Around 50 per cent of the IRDP investments are now made in the secondary and tertiary sectors, based on local resources and local requirements. These include processing industries, handlooms and handicrafts.

Again, in most villages there is scope for tailoring and ready-made garments, chemist shops, woodcraft, country tiles, general store etc. In addition, in villages of a reasonable size, other business/service ventures like flour milling, motor rewinding, cycle repair etc. could also be promoted under IRDP.

Greater emphasis should be placed on developing rural industries which would catalyse the overall development of that area. In a situation of an ever changing pattern of demand, emergence of new products and technologies in a dynamic and growing economy, an attempt would be made to integrate the IRDP activities, particularly those in the Industry Services and Business (ISB) sector with the market. This task would be entrusted to a team of experts/professionals, who would prepare sectoral and micro plans for identifying the thrust areas and activities for working out necessary linkages with other departments/agencies.

The artisans in the rural areas, despite their rich heritage and skills, largely belong to the poverty group. The scheme for Supply of Improved Toolkits to Rural Artisans is directed to this particular target group. In the Ninth Plan, to enable the rural artisans to take advantage of the new opportunities thrown up by the market, there has to be a quantum jump in their skills and productivity to make their activities viable and profitable. It is also necessary to support them with appropriate product designs and training, improved technology on the one hand and professional management and marketing support on the other. In other words, the rural artisans should be serviced through a package approach wherein the distribution of improved toolkits is supplemented with the supply of credit and raw material, marketing support and upgradation of existing technology.

The timely and adequate supply of trade specific toolkits would be given added attention. While deciding the source for supply of toolkits, quality and cost considerations, alongwith the post-delivery services offered by the manufacturer would be of considerable significance. It is important to constantly upgrade the design of the toolkits. For this, research and development would be given due emphasis. States would be advised to take the initiative in developing toolkits in at least two or three artisanal trades. The Department of Science and Technology, the National Small Industries Corporation, the Small Industries Services Institutes and the IITs would interact with the State Governments for development of appropriate technology and designs for improved toolkits for various artisan groups. In the process of designing and manufacturing these trade-specific toolkits, there would be constant consultations and dialogue with groups of artisans and reputed craftsmen in that particular trade. To broadbase the source for supply of toolkits some master craftsmen would also be trained in the manufacture of improved tools/ toolkits. An effort would also be made to develop capabilities for design and upgradation of improved toolkits by artisans themselves. There are pockets of rural technology which have survived the onslaught of modern technological innovations by virtue of their sturdiness and locale-specific utility. These would be identified and replicated elsewhere. For the dissemination of technology the State Governments, in collaboration with the manufacturing agencies, would undertake promotional activities by organising exhibitions at several locations and on important occasions.

Since the 1991 Census data on rural artisans are still not available, the States would be required to conduct a district-wise survey of the total number of rural poor artisans, the number of such artisans provided with improved toolkits and the balance number to be covered. This exercise would facilitate a planned and phased coverage of this target group under IRDP. While individual artisans would continue to be covered with the supply of toolkits in traditional crafts, it is also proposed to cover the rural artisans through a cluster approach and to provide/saturate each cluster with trade-specific toolkits and related inputs in a package. In the implementation of this group cluster approach, due emphasis would be given to the formation of women artisan groups.

Availability of infrastructure facilities is an essential prerequisite for the success of IRDP activities. Substantial investments in programme infrastructure would be ensured, through a larger apportionment of funds, in consonance with the enhanced provisions of 20 per cent for IRDP infrastructure (and 25 per cent in the North Eastern States). There would be a special emphasis on infrastructure created under ISB sector with the setting up of service-cum-facility workshops at convenient places in the rural areas. These could provide common facilities in the use of machines and equipments to the rural artisans as also for repairing electrical gadgets of various types, agricultural implements, automobile parts and articles of common use. These workshops would also be set up in the tribal areas so that the process of initial value-addition to minor forest produce could be undertaken by the tribals themselves. Funding for infrastructure would also include the setting up or upgradation of technology resource centres.

Provision of marketing facilities is an important aspect of infrastructural development. In the Eighth Plan detailed guidelines were issued for setting up District Supply and Marketing Societies (DSMS) with the objective of providing integrated services to IRDP beneficiaries in the cottage and rural industries sector for the supply of raw materials, marketing of surplus products, information on technological upgradation and extension of credit support. Whereas some State Governments have taken the initiative in this regard, by and large the DSMS or similar bodies do not have much of a presence in most areas. In the Ninth Plan alternative strategies would be formulated for developing a suitable marketing infrastructure under IRDP. These would include (i) provision of transport arrangements for carrying IRDP products to rural/urban markets; (ii) introduction of insurance cover to mobile sellers; (iii) provision of better storage facilities; (iv) setting up and revamping of District Supply and Marketing Societies; (v) sale of IRDP products through Khadi & Village Industries Commission (KVIC) outlets, State Emporia etc., besides networking with DRDA showrooms/ markets; (vi) setting up of quality control centres and consultancy centres which the beneficiaries could approach for advice for improving the quality and standard of their products; (vii) involvement of private sector in marketing by adoption of better packaging techniques, design, input, quality control, brand name etc.; and (viii) launching of a suitable advertisement campaign for IRDP products including organisation of exhibitions, melas. Furthermore, the potential of rural haats as a rural marketing outlet for IRDP products would be fully exploited.

There are certain areas in the country which have a very poor banking infrastructure. The areas of North-East and parts of Jammu & Kashmir fall under this category. Attention would be given to evolving innovative strategies and programmes to take care of the unbanked areas.

However, while diversifying the rural economies in high-productivity sectors, provision of adequate training facilities and upgradation of skills would be given primacy. As we have seen, the TRYSEM programme has been the weak link in the overall strategy for self-employment in rural areas. In the Ninth Plan, therefore, training would be made an integral component of IRDP. This integration would reduce to a great extent, the area-skill mismatch as the training would only be imparted in those trades/ occupations which have a market potential. Yet, training would not be made compulsory as there are certain trades/activities under IRDP which do not require this input.

To strengthen the content and design of the training curriculum, the training institutes would have to constantly upgrade their syllabi in tune with the rapid changes in the job market. A basic foundation course would be a critical ingredient of the training curriculum which would make the trainees aware of simple accounting procedures, book keeping techniques and procedures in financial management, information on how to approach the banks and other financial institutions for loans, where to access the latest technology etc. Training would be given a sharper employment focus wherein the training content would be compatible with the area-specific ground realities and would be specifically imparted in those trades and activities which have a market orientation/potential and which would ensure the beneficiary sustained employment.

The quality of training would be improved. Inadequacy of proper infrastructural support has posed as a bottleneck in this effort. While the emphasis would continue to be on imparting training through the established and recognised training institutes like ITIs, Community Polytechnics, Krishi Vigyan Kendras, etc., the training infrastructure in these professionalised training institutes would be suitably strengthened. Special thrust will be given to the creation of training opportunities for women via strengthening of women ITIs, womens wings in general ITIs and women polytechnic. It is also necessary to upgrade the training skills of the trainers in the various government institutions imparting training to IRDP beneficiaries. The existing craft training centres and skill development institutes etc. would be revamped to cater to the needs of the changing situations. In those blocks where there is a concentration of unemployed youth and where there are no reputed training institutes in the vicinity, mini-ITIs should be set up, but only very selectively. Smaller private institutions and craftsmen will be engaged only where institutional support is not available.

In addition, a more effective liaison and interface between the State Governments, DRDAs and the formal/informal private sector/ NGOs is required in order to identify the available employment opportunities in a region and develop training modules accordingly. Efforts would be made to establish a direct linkage between ITIs and industries in the areas where the youth could either be employed directly or could set up ancillaries to cater to the industrial demand. A similar linkage could be established between ITIs and exporters/export houses on the lines similar to gem cutting training in Gujarat. Such linkages are possible in rural hinterlands of towns/urban agglomerations. It is proposed to develop a Management Information System (MIS) through which important training institutes of relevance for IRDP throughout the country are identified and networked. This would give some idea of the areas deficient in the training infrastructure and would indicate where future investment should flow for meeting the training requirements.

With a view to ensuring that the benefits under the programme reach the more vulnerable sections of the society, the Ninth Plan would continue with the assured coverage of atleast 50 per cent for Scheduled Caste and Scheduled Tribe families, 40 per cent for women and 3 per cent for the physically handicapped. Experience in the implementation of IRDP suggests that programmes focussed on women, or in which women have played a dominant role, have performed better. The DWCRA programme has been an excellent vehicle for extending IRDP credit support for women beneficiaries in some States. The group strategy would, thereby, become the main plank for achieving the stipulated reservation for women over the next few years.

In the Ninth Plan social mobilisation will receive a special thrust. Initiative will be taken to build and strengthen the organisations of the poor with the objective of enhancing their capabilities. In this process voluntary organisations would have to play an important role in empowering the poor through advocacy, awareness generation and social mobilisation. As social animators and rural organisers they would help the poor to form SHGs in order to take advantage of the policies and programmes being implemented by the Government for their economic betterment. A voluntary organisation or a technology group could lend support to group activities by ensuring training, technological upgradation and convergence of various schemes. In addition, the DRDAs and PRIs would provide support for capacity building and provide access to credit, technology and markets to SHGs. A cadre of para professionals from within the community would be created to enhance the capability of the SHGs and help the community to access the facilities and services meant for them.

The DRDAs would be restructured in the light of the 73rd Constitutional Amendment Act, which has enhanced their area of operation, commensurate with a larger inflow of funds. The scope of the activities implemented by the DRDAs has enlarged progressively over the years from the sole implementation of rural self-employment programmes to exercising effective coordination and supervision over other Centrally Sponsored Schemes. In the revised format, the DRDAs would work under the supervision and overall control of the Zilla Parishads. The Chairman of the Zilla Parishad will be the Chairman of the DRDA Governing Body and will preside over their meetings. The organisational structure (staffing pattern) of the DRDAs would be suitably revamped and strengthened with the induction of professional cadres and technical experts. The expertise would have to be developed in the fields of credit, technology upgradation, activity-specific training and infrastructural development. To make this implementing agency effective, the staff would have to be given sufficient exposure to the complexities of poverty eradication, human resource development, gender and related issues. The DRDA's agenda for operation in the Ninth Plan, would not be limited only to the achievement of physical targets, but also to ensure the quality of the programme and realisation of intended benefits for the targeted poor beneficiaries.

WAGE EMPLOYMENT PROGRAMMES

The thrust of the Government policy is on assuring at least 100 days of employment per person per year by coordinating all the wage-employment schemes including the State schemes like the Employment Guarantee Scheme in Maharashtra. Now, with the universalisation of the EAS, both JRY and EAS are operating in all the districts/blocks of the country. At present, the JRY is being implemented through the panchayats and the EAS through the administrative apparatus. However, in consonance with the Government policy, the EAS would also be implemented through the Panchayati Raj Institutions (PRIs) in future. There is, therefore, a near complementarity of objectives and there would be a common implementing agency namely the PRIs in the case of both JRY and EAS.

Wage Employment Programmes

- JRY will be confined to the creation of rural infrastructure according to the felt needs of the people at the village level through panchayat.
- The EAS would be the major wage-employment programme.
- In order to make EAS effective in reaching the poor and more backward areas of the country, it is proposed that funds released to States should be made as per incidence of poverty.
- Works can be taken up in blocks where there is concentration of the poor and underemployed, and workers have registered themselves for manual work.

The only major difference is that while, under JRY the allocations are made on the basis of a specified criteria, the EAS is a demand-driven scheme. In the latter case, only initial notional allocations are made and then depending on the demand for supplementary employment and actual utilisation, requests for additional allocations are made by States/Districts. Therefore, in theory there is a good case for merging these two schemes into a single wage-employment programme. Though the merger of the schemes may be operationally expedient, there would be practical constraints. Under the EAS works are taken up at the block level whenever workers register themselves for casual manual employment. Hence, it is possible that no works are taken up in many villages, particularly those which are' economically better off or where the population is sparse. In this case, these villages would be denied funds presently flowing to them under JRY. This would lead to contradictions as JRY has made the villages more independent in so far as they can take up small works according to the felt needs of the community without having to look for funds from the district administration. The JRY has also been successful in taking development to the people with planning and implementation of works according to their felt needs.

Therefore, it is proposed that the JRY would be confined to the creation of rural infrastructure at the village panchayat level. To the extent that the works would be largely labour-intensive, supplementary wage-employment will be generated and this would be monitored without the existing stipulation of a wage-material ratio of 60:40.

The EAS would be the major wage-employment programme. The funds at present flowing to the block and district levels under JRY would also be rediverted to the EAS, augmenting the resources under this scheme. However, certain precautionary measures would have to be taken in order to make the scheme effective in reaching the poor and the more backward areas of the country, where there is a concentration of the poor and underemployed. Towards this end, it is proposed that funds released to States should be made on the basis of the incidence of poverty in order to prevent the better-off States from cornering a larger share of the funds. To elaborate, in progressive States characterised by higher levels of agricultural productivity, higher per capita income and/or high literacy rates there is little real demand for casual manual work on public works. Yet, in these States demand is artificially created as, in the lean agricultural season, labour is willing to work on public works at the prevailing minimum wage rates which are relatively high. In this way, the State's specific real demand is not being addressed and in fact, there is a diversion of resources from the relatively poorer States to the more progressive ones. In such a situation, EAS would cease to be a genuine demand-driven scheme and, therefore, suitable measures would have to be taken to ensure that scarce resources reach the targeted poor and the more backward areas of the country.

Hence, as a second step, the allocations from the States to the districts will also be made on the basis of a predetermined criterion which reflects the relative backwardness of the district. Below the district level, the works can be taken up in blocks where workers have registered themselves for casual manual work. Urbanised blocks would be excluded from the purview of this scheme. No contractors would be involved. This scheme would essentially be implemented by the PRIs at the block level, but with a specific apportionment of funds to the DRDA/ZPs, for taking up inter-block and district level works. Presently, under EAS the funds are earmarked for specific activities in given proportions. Fifty per cent is earmarked for watershed-related activities. It is argued by many that it may not be fair to pre-empt the bulk of the resources for watershed-related activities being implemented in a few villages to the relative neglect of other villages in a block. In the spirit of democratic decentralised planning, the choice of activities must be left to the Panchayat Samities/Zilla Parishads which can select from the various activities/works envisaged in their block/district plans. However, at least 40 per cent of the expenditure would have to be on watershed-related activities in DPAP areas as water conservation through construction of water harvesting structures has become one of the priorities of planning, particularly in the plateau and rocky areas where water tables are falling rapidly and ground water acquifers need recharging. It may be worthwhile to spell out that the works should be labour-intensive with a 60:40 ratio of wages and materials and should lead to the creation of durable assets and rural infrastructure.

In short, EAS will be the single wage-employment programme operating throughout the country, while JRY will be confined to the creation of durable community assets at the village level. So far, under the JRY a certain quantum of funds was earmarked for Special and Innovative Projects. However, experience gained from the implementation of this component of JRY has indicated that most of the projects were not really innovative in nature but merely a replication of the existing programmes on traditional lines. Hence, there is no need to continue a separate scheme for Special and Innovative Projects.

Currently, the funds for IRDP and its allied programmes are shared between the Centre and the States in the ratio of 50:50, whereas the wage-employment programmes such as the JRY and the EAS are funded by the Centre and the States in the ratio of 80:20. It hardly needs to be emphasised that the wage-employment programmes are more popular with the State Governments on account of this cost sharing pattern. As against this, the equal cost sharing formula for self-employment schemes seems to have dampened the enthusiasm of the State Governments towards IRDP and allied programmes. States with difficult resource position find it hard to contribute their share of the outlay. In the Ninth Plan, it is proposed to alter the funding pattern under IRDP and bring it at par with the rural wage-employment programmes. In fact, the cost sharing formula for both IRDP and the wage employment programmes would be revised to a uniform ratio of 75:25 between the Centre and the States.

Million Wells Schemes (MWS)

In view of its importance in providing a source of irrigation to the target group, the MWS would continue in the Ninth Plan to provide for maximisation of agricultural output. Simultaneously, the level of ground water would have to be maintained by the adoption of suitable recharging practices so that the small and marginal farmers can derive maximum benefits from their small holdings.

Until recently, the focus under the MWS was on the creation of employment, with the secondary objective of providing a source of irrigation. This situation has now been modified by delinking the MWS from the JRY and repositioning it as a beneficiary-oriented scheme of irrigation for enhancing agricultural productivity levels of the small and marginal farmers.

The per capita assistance under the MWS for providing a well on the land of the small and marginal farmer ranges approximately from Rs.30,000 to Rs.35,000, though some States have reported a cost of upto Rs.1 lakh. Such a well can irrigate, on an average, one or two hectares. There have been suggestions from some States to permit the construction of field channels in conjunction with the digging of wells. This is particularly significant in view of the fact that water lifted from the wells is conveyed through open unlined earthern channels by most of the farmers, leading to conveyance losses of about 14 to 19 per cent. Necessary control and diversion structures for open line channels may be provided to avoid wastage of water. The channels could consist of structures of soil, cement or even prefabricated cement concrete.

National Social Assistance Programme (NSAP)

The NSAP was introduced as a social security programme for the welfare of the poor households. However, it must be recognised that it is not a poverty alleviation scheme but more in the nature of a welfare programme and hence, has a limited role in the overall strategy of poverty alleviation. Moreover, the financial assistance provided to poor households in the case of old age, death of primary bread winner and maternity is in the nature of transfer payments and therefore, it should really form a part of non plan expenditure. Furthermore, the performance under this programme indicates that it has not been effective in reaching the intended beneficiaries. Moreover, several States have their own schemes of old age pension and maternity benefit for which provisions are made in the States Plans and this should continue to be within their purview, as it is a State subject. Yet, as this programme has now been introduced as a Centrally Sponsored Scheme it would continue in the Ninth Plan but would require a review thereafter.

Land Reforms

In the Ninth Plan the issue of agrarian restructuring will continue to receive the top most priority in the expectation that the States would be able to facilitate changes that would make for more efficient agriculture, leading to increases in both output and employment. This process will, in turn, contribute to the achievement of a higher rate of economic growth with social justice.

The main components of the land reform policy are the detection of ceiling surplus land and the distribution of the existing surplus land, besides tenancy reform, consolidation of holdings, providing access to the poor on common lands and wastelands, preventing the alienation of tribal lands and providing land rights to women. However, for the successful implementation of land reforms, updating of the land records, both by traditional methods and through computerisation, is an essential prerequisite. Let us elaborate.

Ceiling on Land Holdings: With the introduction of the Land Ceilings Act in 1972, the ceiling on land holdings was introduced in almost all the States with the exception of some North-Eastern States, though the ceiling limit varied depending on the quality of the land. The ceiling surplus land was to be distributed among the landless poor. In this way, land ceiling was considered an important instrument for reducing disparities in the ownership of land and as a way of increasing productivity through greater utilisation of labour. However, in practice, the extent of land declared surplus was very limited. The total area declared surplus is estimated at less than 2 per cent of the total cultivated area. Further, distribution of surplus land has been limited in several States because of institutional and legal rigidities. There are other reasons too which have led to the poor performance with regard to land ceiling. Most importantly, the ceiling surplus land continues to exist in a concealed way particularly in areas which have been covered by irrigation in the post-1972 era. Further evidence suggests that there are areas where land owners have land in excess of ceiling limits which can be mopped up if the programme for unearthing it is pursued vigorously. Also, in order to circumvent the ceiling laws, benami and furzi transfers are effected which need to be identified. There is clearly a need to detect the land falling within the ceiling limits and redistribute it. But detection and distribution of ceiling surplus land require tremendous political and administrative will.

In the wake of the economic liberalisation in certain States like Karnataka, the industry and the large farmers are being given exemption from ceiling laws without seeking the permission of the Government of India. This would certainly go against the interest of the poor as it would increase landlessness and depress agricultural wages. Hence, this issue requires close examination, before such exemptions are given.

Small farms are efficient in terms of yields and returns per unit of area. However, they are often not able to generate adequate income to sustain further investment. Therefore, there is a need for both horizontal and vertical diversification of small farms. An analysis of Indian Council of Agricultural Research (ICAR's) data of national demonstrations shows that there are large technological/ yield gaps, particularly in the backward regions, which, if fully exploited, would enable the marginal and small farmers to generate incomes well above the sustenance level, subject to crop rotation, cropping pattern, yield responses of technological adoption, etc. In fact, in irrigated areas, the small and marginal farmers are viable because of their capacity to produce two or three crops a year. In the case of rainfed agriculture, a breakthrough in dryland farming is required with respect to the traditional cereal crop that can grow in these areas.

Alternatively, there is a possibility of shifting to other high-value crops like fruits, vegetables, mulberry, etc. Other support systems would have to be appropriately developed to provide facilities for credit, marketing, storage and transportation of the perishable high-value crops. In fact, a change in the cropping pattern away from cereals to some high-value crops, would certainly make the small and marginal farmers viable. Also, given the increasing market demand for certain products, sericulture, horticulture and aquaculture have a great potential. In Maharashtra and West Bengal a small piece of land yields adequate returns from horticulture and acquaculture respectively. This provides further justification for ensuring that ceilings are strictly enforced as small farms can be perfectly viable in the given context.

As observed in the UNDP Human Development Report 1996, as land is redistributed from big to small farms, not only the family labour per hectare can increase sharply, so, can hired labour also. For both the reasons, the employment situation improves even for those, who remain landless after the land reform.

The main conclusion from this is that an agricultural strategy centered on small farms, rather than large, simultaneously increases the social efficiency of resource use in agriculture and improves social equity through employment creation and more equal income distribution, that small farms generate.

While there appears to be a rationale for reducing the existing ceiling limits further with a view to alleviating poverty of the growing number of rural landless poor, this is unlikely to have the support of the State Governments. Therefore, a pragmatic approach would be to strictly enforce the existing ceiling limit without permitting any attempt to circumvent it. Given the employment elasticity of agriculture in areas, which are agriculturally backward, land reforms would ensure both agricultural growth and greater employment for the rural poor. It would also provide a social status to the large number of poor. A greater transparency in the method of distribution of surplus land is possible with a greater involvement of panchayats, local communities and NGOs.

Tenancy Reforms: The policy with regard to tenancy clearly provided for conferment of ownership rights on tenants or for acquisition of ownership rights by them on payment of a reasonable compensation to the landlords. The tenancy laws of most States abolished tenancies so as to vest ownership of land with the actual tiller. Despite the legal provisions to abolish tenancy, it has continued to flourish with the existence of a large number of tenants and sub-tenants without any protection against eviction leading to insecurity among them. The continued existence of oral and concealed tenancies has led to low investments and low productivity in agriculture in several States where the implementation of tenancy reforms has been tardy. In contrast, in West Bengal, Karnataka, and Kerala, much success has been achieved in this respect. The success of 'Operation Barga' in West Bengal is well documented. This shows that conferring occupancy rights on the tenants has led to better investments in land and consequently, a higher rate of return. Hence, it is desirable that States which have a high incidence of concealed tenancy recognise this fact and record the rights of tenants. In addition, absentee landlordism has to be restrained. To the extent that large land owners leave their land fallow, either the State should take it over and lease it out on a long-term basis or the land owners should be required to lease it out on a long-term basis. Measures would have to be taken for the protection of tenants against displacement, eviction and other forms of exploitation.

A ban on tenancy was imposed in almost all the States to encourage owner cultivation and to give security of tenure to the sharecroppers and the tenants. In areas characterised by semi-feudal modes of production, where agricultural markets are not well developed, this ban is desirable with a view to protecting the tenants. It has been demonstrated that in 1972-73 in Purnea district of Bihar 40 per cent of the land was under sharecropping. This percentage has gone down and is probably around 25 per cent. The share of landholders is limited to 25 per cent of the gross produce but in practice this is perhaps illusory. Since the tenancies are oral and the sharecroppers are weak their hold on land is tenuous and they have to give to the landlord more than half of the produce. In this case, the tenants have no incentive to make long-term investments in land. The landlords also do not look upon it as a productive asset but a store of value and a reflection of their social status. Both the landlord and the sharecropper would gain if the sharecropper had the rights of cultivation in the land as he would make greater investments which would lead to higher returns. This shows that in areas where tenancy flourishes, reforms are required together with a proper implementation of laws.

On the other hand, in the "Green Revolution" areas where there is greater awareness and the markets work, freeing the lease market for land would contribute both to equity and efficiency. In these areas where the traditional arrangements exist, the tenants have been reduced to the status of an agricultural labourer with the landlord exercising considerable influence. In this case, the tenants as well as small and marginal farmers would be able to augment their operational holdings by leasing in land and, with a greater intensity of cultivation, it would lead to greater output. Clearly, agricultural tenancy should be opened up and leasing in of land permitted subject to the ceiling limits. This would activise the land market which would enhance the poor people's access to land.

Protection of Tribal Land: Despite the commitment that the tribal lands must remain with the tribals, alienation of the tribals from their land continues on a large scale due to various legal loopholes and administrative lapses. Hence, in the Ninth Plan legal provisions must be made for the prevention of alienation of tribal lands and for their restoration, not only in the notified scheduled tribe areas but also in the tribal lands in other areas. Also, the regulation of resale of the tribal lands should be made as stringent as possible. Given that alienation is basically a consequence of economic deprivation and social discrimination it is felt, that for the tribal communities, various development programmes must be dovetailed in order to improve the income level of these people and to provide them with a basis for sustained livelihood. Even in the case of acquisition of lands of the tribals by the Government for public purposes, it should be restricted to a minimum. Encroachment of land belonging to tribals should not be permitted. Even civil courts should have no jurisdiction in the proceedings involving transfer of land of persons belonging to scheduled tribes.

Consolidation of Holdings: As already stated, consolidation of holdings has been successful in a very few States though several States have enacted legislation in this regard. In some cases, there are genuine problems in the process of consolidation including proper valuation of land, fear of eviction of small and marginal farmers who are tenants, inadequate availability of staff and lack of updated land records. Despite these constraints, consolidation of holdings makes for efficient land use and water management, leading to higher productivity and, therefore, must be enforced wherever practicable. However, in so far as there is insecurity among the tenants and small and marginal farmers, the State must ensure that their interests are protected. Consolidation operations, whenever undertaken, should be integrated with survey and settlement operations in order to avoid duplication in work and harassment to the affected person. The involvement of the Panchayati Raj Institutions should facilitate this process through a greater participation of the village people.

Government Land and Common Property Resources: In rural areas every village has common lands as well as other common property resources. These are a source of sustenance to the landless. Evidence suggests that considerable area of the government land has been taken over both by the rural poor as well as by the rural elite for agricultural and housing purposes. Unluckily, due to negligence, unsustainable overuse and excessive pressure of population on land, the productive capacity of common lands has been diminishing. However, joint forest management and watershed development programmes are schemes which can be successful in the regeneration of these common lands. Clearly, 'pattas' should be given to the rural poor in order to provide them access to a means of livelihood. There are several success stories, both in the area of joint forest management as well as watershed development by groups of people, especially women on common land, which could be replicated. These lands can also be used for providing grazing land, fodder and fuel to the poor. If the common property resources vest in the Gram Panchayat, access to it should be limited to the rural poor and the rural elite should not be allowed to encroach upon it. Similarly, the wastelands which are lying unutilised should be reclaimed and distributed among the marginal farmers and landless labourers of the area. This would provide them a source of income as also generate greater output from a variety of activities in the area of agriculture, horticulture, fodder, fuelwood and other agro-forestry. Given the paucity of 'cultivable' land, it is necessary to redistribute the wastelands and fallow land in order to provide productive employment to the poor.

In the case of land vested and in possession of the Government and/ or where a final decision is pending in a judicial court, it is suggested that such land should not be kept vacant but allotted to eligible rural poor on a short-term basis with a clear understanding that they would have to handover the land to the original landlord if the court so directs. However, legal opinion is being sought on the possibility of taking up this measure. Again in the case of Government wasteland and degraded forests, there is a growing opinion that such land should be allotted/leased to the corporate sector and for industrial and commercial purposes. As argued above, this too would adversely affect the rural poor. Access to such land should be restricted to the rural poor in order to provide them with employment.

Gender and Land Rights: So far, the strategy of land reforms has not given any cognizance to the existing gender inequalities in land inheritance laws and ceiling laws. In most regions of the country, women constitute a disproportionate number of poor. They are also more dependent on agriculture for a livelihood than men, as men shift to non-farm employment. Also, it is estimated that 20 per cent of the rural households are *de facto* female-headed. Yet, very few women have titles to land and even fewer control it. Hence, ensuring women's effective command over land will be one of the new priorities of the Ninth Plan.

Traditionally, it was accepted that agricultural land would be inherited by sons, even though in some States the inheritance law did not stipulate such a provision. Hence, it is necessary to implement the laws and also to record the rights of women, where it is legally claimed by them. Further, in so far as inheritance rights in tenancy laws are subject to specific land acts, the issues cannot be easily resolved. In some States the tenancy laws provide for devolution of tenancy to male descendants and only in their absence, women can inherit, but then also to a limited extent. Hence, it is recommended that States should amend their land and inheritance laws so as to bring agricultural land at par with other forms of property. There has been some progress in this direction in Southern and Central India.

Many of our States have improved women's access to land and landed property. States like Karnataka, Tamil Nadu and Andhra Pradesh have amended the Hindu Succession Act, 1956 to formalise issues related to women's right to property including land. Some, like Rajasthan and Madhya Pradesh have provided that issues related to property, including landed property, would be dealt with in accordance with the appropriate personal laws. However, serious anomalies continue to persist. A number of States like U.P. Haryana, J&K, Delhi and Punjab are apparently yet to take adequate steps to provide the Constitutional/legal safeguards to women with respect to their access to land.

Additionally, it is necessary that when ceilings are fixed in relation to a "family unit", the definition of a family should be uniform across States and sons and daughters should be given equal consideration, both while assessing ceiling surplus land and in land distribution under various resettlement programmes. The pattas should be in the name of women to a larger extent. While joint pattas are better than no pattas, which do not provide the women control over it. In fact, groups of poor rural women should be given group pattas with usufructuary rights, but no right to sell individually. This group approach would enable the women to retain control over land. They could invest in the land collectively and cooperate in sharing both labour time and the returns. There are several instances of such joint management by groups of women which need to be replicated. However, to enable women to reap the benefits of land acquisition, greater access to information, credit, inputs, marketing and technologies must be provided to them.

In order to have a better data base, both the Agricultural Census and the National Sample Surveys should provide information on land ownership, land operated and land under tenancy by gender. So far, land-based statistics are recorded on a household basis. A pilot survey should be undertaken proceeding the next nationwide NSS data collection exercise.

Updating of Access to Land Records: Maintenance of up-to-date land records is crucial for effective implementation of land reforms. In recognition of this, during the Seventh Plan period, a Centrally Sponsored Scheme for Strengthening of Revenue Administration and Updating of Land Records was introduced with a view to strengthening survey and settlement organisations at the grassroots level through training, equipment, staff etc. Hence, it is imperative that States should give special emphasis to resurveys and adopt appropriate modern technologies for this, including aerial survey, photogrametic systems, global positioning system, use of scanners, digital computerised maps etc.

Further, a special drive for recording land rights of tenants and sharecroppers, as in West Bengal, must be undertaken by other States. As recommended by the Conference of Revenue Ministers held in 1997 all the existing tenancy reform legislations should be examined and a model legislation prepared.

Another Centrally Sponsored Scheme for Computerisation of Land Records was started during 1988-89 as a pilot project. But the scheme has made little progress. In fact, several factors including power shortage, lack of trained staff and bureaucratic procedures involved at various stages, have constrained the progress of computerisation of land records. Despite these problems, computerisation must be given a high priority and solutions found for the operational problems.

The updating of land records can be expedited even without computerisation through the involvement of panchayat institutions and the local revenue functionaries. The village-level revenue functionaries should be placed under the control of the Gram Panchayats, though the appellate jurisdiction should continue with the tehsildar. At the district level, the land revenue system must work under the Zilla Parishad. The 30 per cent representation for women in PRIs should help the cause of women in so far as recording of rights of women in land is concerned.

Moreover, steps have to be taken to bring about greater transparency in the administration of land records, with access to information regarding land holdings on demand by any individual. Copies of land records, including record of rights, field books/ field map, Land Pass Books as also mutation statements, status of land and jamabandi register should be accessible and copies provided on payment of a fee.

ANTI-POOR LAWS AND POLICIES

Over time, the expenditure on the various poverty alleviation programmes has increased significantly. It is true that these programmes play a vital role in ensuring that the poor are able to sustain themselves, particularly during the lean seasons, through generation of additional employment and incomes. However, it is becoming increasingly clear that spending money is not the only way of ameliorating the conditions of the poor. Scant attention has been paid to the role of non-monetary policies and institutional arrangements which affect the lives of the common people, especially the poor. There are several laws and policies which are anti-poor. The knowledge about them is scattered and often anecdotal. However, some examples may illustrate the nature of the problem.

For tribals and forest dwellers minor forest produce obtained from forest and common property resources is a major source of livelihood. However, the poor tribals are facing problems in gathering this minor forest produce due to diversion of non-timber forest produce to the industry. Bamboo forests are being readily leased to the paper industries regardless of the provision in the Forest Policy 1988 which stipulates that the forest dwellers have the first charge on the forest produce. Rural artisans, who make products out of bamboo, face shortage of raw material as the forest departments allot them green bamboo in rationed quantities. This practice has been observed in Karnataka and Orissa. In some cases, forest land is leased out to private industry for long periods. This too adversely affects the poor.

The forest department also issues licences to women for gum collection from babool trees and compels them to sell their entire collection to the forest department at a prescribed price which is one third to one fourth of the market price. In some States, tribals can collect hill brooms but cannot process these into a broom nor can they sell the collected items in the open market. Thus, the poor are prevented from making value addition through processing and are denied the right to get the best price for their produce. Licensing activities for the poor in rural areas are cumbersome and time consuming. To set up a charcol kiln one requires four to five types of permission.

The cattle flayers in UP have no legal control over their own produce i.e. the hides they flay from the naturally fallen animals. The Zilla Parishads award contracts either to individuals or to cooperatives for collection and storage of hides. Hence, the flayers are not able to sell in the open market but are forced to accept the terms made by the contractors. Again in Bihar, auction ferry right is made in favour of Zamindars and contractors who collect tolls and taxes from the poor who have to cross the river to sell their produce. In addition, those who are fishing or plying boats across the river have to pay taxes to the Zamindars. Normally, access to, and use of, the river should be free to farmers, labourers, petty traders etc. In Assam, the fishery laws are such that the fishermen do not benefit from them. A sector of the river is leased out to "Mohildhars" who auction the fish and only 1/3rd is given to fishermen who actually catch the fish. These are ways in which common property resources are being used in a way that are prejudicial to the poor.

The rural poor do not have a level playing field as the market for their finished products are dominated by a single trader or cartel of traders operating in the area. In the market, the terms of trade are against them. Neither the public nor the private sector has encouraged the growth of poor people's organisations but, instead, has either tried to suppress or control them. Cooperatives could have become the real supporters of the poor and needy but the process for their formation and registration is very cumbersome. Where cooperatives of poor people have been successful such as womens' beedi cooperatives, they have been taken over by private companies. Similarly, the Government has often interfered with the independence of the cooperatives by putting government operators incharge of them.

Such examples abound. An exercise has already been done to document some of these laws and policies. In the Ninth Plan, an attempt will be made to initiate the process of identifying anti-poor laws/policies, Statewise. These would be brought to the notice of the policy makers, local governments and NGOs so that these may be suitably modified and/or repealed in the interest of the poor.

INTEGRATION OF POVERTY ALLEVIATION PROGRAMMES WITH SECTORAL PROGRAMMES

It is necessary to recast the special employment programmes with a view to making them more effective in meeting not only the short-term objective of providing temporary work, but also in building up the productive capacity of individuals/areas which, in turn, would make for greater employment on a more sustainable basis. The focus would have to be on agriculture and allied activities, besides rural non-farm sectors and services, which have a high employment elasticity. This would require a high degree of convergence among the various poverty alleviation programmes (PAPs), area development programmes and sectoral schemes, within a district plan based on the physical and human endowments of the area, the felt needs of the people and the total financial resources available. Using scientific methods, remote-sensing agencies at the Centre and State level would be asked to provide detailed maps showing land, water and other physical resources of the area, with the aid of photogrammetry and satellite imagery. The detailed maps would then be scrutinised to identify all possible watersheds. Planning along watershed lines would ensure minimum surface run-off, thus conserving water from rainfall. Viable activities in agriculture and allied sectors would have to be selected. Agro-processing activities linked with the cropping pattern, village and small industries with growth potential and other infrastructural gaps would also have to be identified and prioritised. Within this framework, poverty alleviation programmes would be integrated with area development programmes within a developmental plan at the district level.

So far, there has been a complete dichotomy between various sectoral as well as poverty alleviation programmes that have been planned and implemented by the concerned line departments. The Government has recognised this dichotomy but greater efforts have to be made to effect the convergence in practice. As a start, an attempt has been made to integrate DPAP and DDP, EAS and the Integrated Wasteland Development Programme (IWDP), all being implemented by a single Ministry. Watersheds are to be constructed and associated works of drainage, land development and terracing undertaken. Also, afforestation, agro-based and horticultural development, pasture development, crop demonstration for popularising new crops/varieties and upgrading of common property resources are being taken up. Clearly, there is scope for integrating other sectoral programmes of soil and water conservation, forestry, minor irrigation, animal husbandry, agriculture and other departments, funds from which flow from sectoral heads to the district level. Once the area plans are prepared, dovetailing of funds would not pose a problem.

Institutional Mechanism for Delivery

Consequent to the 73rd Constitutional Amendment Act, State Governments have enacted enabling legislations providing for local bodies at the village, intermediate and district levels. In almost all the States with the exception of Bihar the PRIs have been duly constituted. The State Governments are required to endow these PRIs with the requisite financial and administrative powers to enable them to function as institutions of self-government. The PRIs would be responsible for preparing the plans for economic development and social justice through the District Planning Committee and for implementing them, in respect of the items listed in the Eleventh Schedule of the Constitutional Amendment. This process of devolution is at various levels of operationalisation for individual States. It is expected, that in the Ninth Plan, the States would devolve funds on the panchayats both from the Consolidated Funds of the States and the allocations made by the Central Government for Centrally Sponsored Schemes. In addition, the panchayats have to be given their own revenue raising powers, as per the recommendations of the State Finance Commissions. It is expected that village level plans would be prepared, based on the felt needs of the people as articulated in the Gram Sabha meetings. These plans would be incorporated into the intermediate level plans and finally merged into a district plan. This district plan would then enable the dovetailing of funds from the various sectoral poverty alleviation and area development programmes. In this way, development planning would begin from below reflecting the aspirations of the people within the constraints of the available physical and financial resources.

Voluntary organisations would also play an enhanced role especially as facilitators and social animators in bringing about greater awareness through advocacy. They would also help the poor to form self-help groups with the objective of improving their economic status through concerted action. In this way the PRIs, the voluntary organisations and the community would work in tandem to bring about greater development at the local level and consequent reduction in poverty levels.

URBAN POVERTY ALLEVIATION

Poverty reduction is an important goal of the urban policy. Urban growth is a result of (1) natural increase in population (2) net migration from rural areas to urban areas and (3) reclassification of towns. The common notion that migration largely fuels urban growth is only partially correct. Therefore, it is necessary to view urban poverty as distinct from rural poverty and not as mere transfer of rural poverty into urban areas.

Urban poverty leads to : (a) proliferation of slums and bustees; (b) fast growth of the informal sector; (c) increasing casualisation of labour; (d) increasing pressure on civic services; (e) increasing educational deprivation and health contingencies.

The Urban Poverty Alleviation Programmes (UPAPs) which were in operation during eighth plan are as follows:

Nehru Rozgar Yojana (NRY)

In order to alleviate the conditions of urban poor, a Centrally Sponsored programme-Nehru Rozgar Yojana-was launched at the end of the Seventh Five Year Plan (October 1989) with the objective of providing of employment to the urban unemployed and underemployed poor. The Central Government indicated its overall contribution while the essential task of identifying, earmarking and coordinating the relevant sectoral inputs was undertaken by the State Governments. The NRY consisted of three schemes namely (i) the Scheme of Urban Micro Enterprises (SUME); (ii) the Scheme of Urban Wage Employment (SUWE); and (iii) the Scheme of Housing and Shelter Upgradation (SHASU). The physical and financial details of the implementation of these Schemes are at Annexure-I and Annexure-II respectively. During the Eighth Plan, 92% of the available funds were utilised and but for the shortfall in the number of dwelling units upgraded/in progress under SHASU, the targets have been achieved under all the other schemes.

Urban Basic Services for the Poor (UBSP)

The UBSP Programme was implemented as a Centrally Sponsored Scheme during the Eighth Five Year Plan with the specific objectives of effective achievement of the social sector goals; community organisation, mobilisation and empowerment; and converence through sustainable support system. The expenditure on the Programme was being shared on a 60:40 basis between the Central and the State Governments and UTs (with legislatures). Further, the per capita expenditure on any slum pocket is Rs.75/-in the first year and Rs.50/-from the second year onwards after the basic infrastructure is developed. The UBSP was targetted to cover 70 lakh urban poor beneficiaries in 500 towns during the Eighth Plan period. The Programme has achieved the physical target of 70 lakh beneficiaries during the Eighth Plan period in 350 towns. Against the release of the Central share of Rs.8090 lakh, the release of the State share was Rs.3439.64 lakh. As on 31.03.1997, 353 towns and 4993 slum pockets have been selected for coverage and 75 lakh beneficiaries have been covered.

Prime Minister's Integrated Urban Poverty Eradication Programme (PM IUPEP) :

Recognising the seriousness and complexity of urban poverty problems, especially in the small towns where the situtation is more grave due to lack of resources for planning their environment and development, the PMI UPEP was launched in November, 1995. The PM IUPEP was a Rs.800 crore scheme approved for the period up to the year 2000. Programme was applicable to all Class II urban agglomerations with a population ranging between 50,000 and 1 lakh subject to the condition that elections to local bodies have been held. The Programme was being implemented on a wholetown/ project basis extending the coverage to all the targetted groups for recuring a visible impact. During 1995-96 and 1996-97, Rs.176.40 crore were released to the States/UTs. Most of the States are in the preparatory stages of the Programme, such as house-to-house survey, patial mapping, need assessment, developing lternative project reports, building community structures etc., which take quite some time. The physical achievements as reported by the States are as under:

a. House-to-house survey has been completed in 213 towns.

b. Town-wise project reports have been prepared for 229 towns.

c. Under the self-employment component, 20775 applications have been forwarded to banks, out of which 3080 cases have been approved.

d. Under the Shelter Upgradation Component, 10386 applications have been forwarded to banks/HUDCO, out of which 4743 cases have been approved by HUDCO.

e. As many as 8382 Neighbourhood Groups, 1200 Neighbourhood Development Committees and 444 Thrift and Credit Societies have been formed.

The performance of the UPAPs in the Eighth Five Year Plan reveals:

- that although urban poverty is no less severe than rural poverty, the priority accorded to alleviation of urban poverty is low as the common perception is that urban poverty is a transfer of rural poverty into urban areas;
- that UPAPs are highly fragmented and have overlapping objectives and strategies.
- that integration of UPAPs with sectoral development and area development programmes has been overlooked.
- that the role of voluntary organisations and community based organisation in planning and implementation of UPAPs is on the periphery.

The Planning Commission's (Modified Expert Group) estimates of urban poverty show a decline in the percentage of urban poor from 38 percent (1987-88) to 32.36 percent (1993-94). As per the Modified Expert Group the number of urban poor is 763.37 lakh in 1993-94. The Hashim Committee, which was set up to review and rationalise Centrally Sponsored Schemes for poverty alleviation and employment generation, had gone into the question of rationalisation of the existing poverty alleviation programmes and recommended that:-

1. the self-employment component of NRY and PMIUPEP be combined into a single programme valid for all the urban areas all over the country;

- The Planning Commission's (modified Expert Group) estimates of urban poverty show a decline in the percentage of urban poor from 38 percent (1987-88) to 32.36 percent (1993-94). As per the Modified Expert Group the number of urban poor is 763.37 lakh in 1993-94.
- The Hashim Committee, which was set up to review and rationalise Centrally Sponsored Schemes for poverty alleviation and employment generation, had gone into the question of rationalisation of the existing poverty alleviation programmes.
- Based on the recommendation of the Committee the Swarna Jayanti Shahari Rozgar Yojana (SJSRY) has been launched with effect from 1.12.97 and NRY, PMIUPEP and UBSP have been phased out.
- The programme has two sub-schemes namely, (a) Urban Self –Employment Programme and (b) Urban Wage Employment Programme. The self-employment and wage employment components of the NRY and PMIUPEP have been re-organised under this single programme. The shelter upgradation components of both NRY and PMIUPEP has been merged with the National Slum Development Programme.

2. the urban wage employment component as well as the physical infrastructure development component under the NRY and the PMIUPEP be merged and be made applicable to all the urban areas with a population less than 5 lakhs. This component may be separated from the self-employment component as a separate scheme with a distinct identity; and

3. the shelter upgradation/housing component under NRY and PMIUPEP be retained either as a separate scheme or merged with the Slum Development/Basic Services Schemes operating at present.

The Swarna Jayanti Shahari Rozgar Yojana (SJSRY)

In pursuance of the above recommendations, during the Ninth Plan it is proposed to launch the Swarna Jayanti Shahari Rozgar Yojana (SJSRY) and phase out NRY, PMIUPEP and UBSP. The SJSRY is to be a Centrally Sponsored Scheme applicable to all the urban areas with expenditure to be shared in ratio 75:25 between the Centre and States/UTs.

The programme has become operational on December 1, 1997. This programme would have two sub-schemes, namely, (i) UrbanSelf-Employment Programme and (ii) Urban Wage Employment Programme. The self-employment and wage employment components of the existing NRY and PMIUPEP have been reorganised under this single programme. Further,the shelter upgradation components of both NRY and PMIUPEP will be merged with the National Slum Development Programme.

The Swarna Jayanti Shahari Rozgar Yojana (SJSRY) seek to provide gainful employment to the urban unemployed or underemployed poor by encouraging the setting up of self-employment ventures or provision of wage employment. This programme will rely on the creation of suitable community structures on the UBSP pattern and delivery of inputs under this programme will be through the medium of urban local bodies and similar community institutional structures.

The Swarna Jayanti Shahari Rozgar Yojana rests on the foundation of community empowerment. Towards this end, community organisations like Neighbourhood Groups (NHGs), Neighbourhood Committees (NHCs) and Community Development Societies (CDSs) will be set up in the target areas based on the UBSP pattern. The CDSs will be the focal point for purposes of identification of beneficiaries, preparation of applications, monitoring of recovery and generally providing whatever other support is necessary to the programme. The CDSs will also identify viable projects suitable for that particular area.

These CDSs may also set themselves up as Thrift and Credit Societies to encourage community savings, as also other group activities. A maximum expenditure at the rate of Rs.100 per member for the first year and Rs.75 per member for each subsequent year will be allowed for activities connected with the CDSs.

The Urban Self Employment Programme (USEP):

This programme will have three distinct components:-

i. assistance to individual urban poor beneficiaries for setting up gainful self-employment ventures.

ii. assistance to groups of urban poor women for setting up gainful self-employment ventures. This sub-scheme may be called "The Scheme for Development of Women and Children in the Urban Areas (DWCUA)".

iii. training of beneficiaries, potential beneficiaries and other persons associated with the urban employment programme for upgradation and acquisition of vocational and entrepreneurial skills.

Coverage:

i. The programme will be applicable to all urban towns in India.

ii. The programme will be implemented on a whole town basis with special emphasis on urban poor clusters.

Target Groups:

i. The programme will target the urban poor,i.e those living below the urban poverty line, as defined from time to time.

ii. Special attention will be given to women, persons belonging to Scheduled Castes/Tribes, disabled persons and other such categories as may be indicated by the Government from time to time. The percentage of women beneficiaries under this programme will not be less than 30 per cent. The SCs and STs must be benefited at least to the extent of their proportion in the local population. A special provision of 3 percent will be reserved for the disabled under this programme.

iii. There will not be any minimum educational qualification set for beneficiaries under this programme. However, to avoid an overlap with the PMRY scheme, for self-employment component, this scheme will not apply to beneficiaries educated beyond the IX standard. As regards the wage employment component, there will be no restrictions of educational qualifications whatsoever.

iv. A house-to-house survey for identification of genuine beneficiaries will be undertaken. Non-economic parameters will also be applied to identify the urban poor in addition to the economic criteria of the urban poverty line.

All other conditions being equal, women beneficiaries belonging to women-headed households will be ranked higher in priority than other beneficiaries.

COMPONENTS

(i) Self-employment through setting up Micro-enterprises and Skill development:

To avoid duplication with the ongoing Prime Minister's Rozgar Yojana (PMRY), this component of SJSRY is confined to the below poverty line beneficiaries who have no education upto ninth standard, with emphasis on those accorded a higher priority on the basis of the non-economic critera. The maximum unit cost will be Rs.50,000 and the maximum allowable subsidy will be 15 percent of the project cost, subject to a limit of Rs.7,500. The beneficiary is required to contribute 5 percent of the project cost as margin money. In case a number of beneficiaries decide to jointly set up a project, such project will be eligible for a subsidy which will be equal to the total of the permitted subsidy per person as per the above criteria. In this case too the provision relating to 5 percent margin money per beneficiary will apply. The overall project cost, which can be permitted, will be the total of the individual project cost allowable per beneficiary.

Skill development through appropriate training is another element of this programme. The unit cost allowed for training will be Rs.2000 per trainee, including material cost, trainers' fees, other miscellaneous expenses to be incurred by the training institution and the monthly stipend to be paid to the trainee. The total training period for skill upgradation may vary from two to six months, subject to a minimum of 300 hours. Infrastructural support may also be provided to the beneficiaries setting up micro-enterprises in relation to marketing of their products etc.

(ii) Development of Women and Children in Urban Areas (DWCUA):

This scheme is distinguished by the special incentive extended to urban poor women who decide to set up self-employment ventures as a group as opposed to individual effort. Groups of urban poor women will take up an economic activity suited to their skill, training, aptitude and local conditions. Besides generation of income, this group strategy will strive to empower the urban poor women by making them independent as also providing a facilitating atmosphere for self-employment. To be eligible for subsidy under this scheme, the DWCUA group should consist of at least 10 urban poor women.

Financial Pattern

The DWCUA group society will be entitled to a subsidy of Rs.1,25,000 or 50% of the cost of project whichever is less. Where the DWCUA group sets itself up as a Thrift and Credit Society, in addition to its other entrepreneurial activity, the group/Thrift and Credit society will also be entitled to a lump sum grant of Rs.25,000 as a revolving fund at the rate of Rs.1000 maximum per member. This revolving fund will be available to a simple Thrift and Credit Society also even if the society is not engaged in any project activity under DWCUA.

The Urban Wage Employment Programme (UWEP):

This programme seeks to provide wage employment to beneficiaries living below the poverty line within the jurisdiction of urban local bodies by utilising their labour for construction of socially and economically useful public assets. This programme will apply to urban local bodies, the population of which is less than 5 lakhs as per the 1991 Census.

The material-labour ratio for works under this programme will be maintained at 60:40. The prevailing minimum wage rate, as notified from time to time for each area, will be paid to the beneficiaries under this programme.

This programme will be dovetailed with the State sector EIUS scheme as well as the National Slum Development Programme (NSDP). This programme is not designed to either replace or substitute the Environmental Improvement Of Urban Slums (EIUS), the NSDP, or any other State sector schemes.

Project Administration :

At the community level, a Community Organiser(CO) will be appointed for about 2000 identified families. The Community Organiser should, as far as practicable, be a woman.

At the town level, there will be an Urban Poverty Eradication Cell under the charge of a Project Officer. The Project Officer will be responsible for coordinating the activities of all the CDSs and COs.

At the district level, the State Government will constitute a District Urban Development Agency (DUDA) with an officer designated as the District Project Officer.

At the State level, there will be a State Urban Development Authority (SUDA), which will be headed by a full-time senior officer of the State Government. The SUDA will be designated as the State Nodal Agency for urban anti poverty programmes.

At the national level, the Department of Urban Employment and Poverty Alleviation will be the nodal department. The programme will be monitored and overseen by the UPA Division.

The Ninth Plan, in particular, needs to focus on the effective implementation and management of urban poverty alleviation programmes (UPAPs). Targetting of urban poor is essential for the success of any anti-poverty programme. Therefore, the UPAPs should be designed to accomodate flexibility in operation. The key components of the Ninth Plan strategy would be:

- effective targetting using either a self-identification system or a selection process by institutions responsible for financing and managing UPAPs as faulty targetting will enhance income inequalities.

- a "bottom up" approach should be adopted for UPAPs which should be routed through community based organisations. There should be effective participation of community institutions for extending loans and for aiding loan recoveries made under UPAPs.
- since the problem of urban poverty is a manifestation of the higher incidence of marginal and low income employment in the informal sector, it is essential to upgrade informal sector occupations. Hence there is a need for emphasis on generation of self-employment in processing and services sector, improving the acccess to technology and credit and above all improving the general legal and physical environment which governs the working of the informal sector.

The policy framework adopted by the Government of India also mentions that "a frontal attack on poverty is an important element in development policy. This is the main rationale for anti-poverty programmes".

Bibliography

Ausubel, F.M.: *Current Protocols in Molecular Biology*, New York, John Wiley and Sons, 1989.

Brandwein, P.F.: *Sourcebook for the Biological Sciences*, San Diego, Harcourt Brace JOvanovich, 1986.

Brooks, Rodney A.: *Flesh and Machines: How Robots Will Change Us*, New York, Pantheon Books, 2002.

Chrispeels, Maarten : *Plants, Genes and Crop Biotechnology*, Sudbury MA, Jones and Barlett Publishers, 2003.

Cox, Earl D. : *Beyond Humanity: Cyber Evolution and Future Minds*, Roackland, Charles River Media, 1996.

David Sadava: *Plants, Genes and Crop Biotechnology*, Sudbury MA, Jones and Barlett Publishers, 2003.

Dixon, Dougal: *After Man-A Zoology of the Future*, New York, St. Martin's Press, 1981.

Dodds, John H.: *Plant Genetic Engineering*, New York, Cambridge University Press, 1985.

Drexler, K. Eric: *Engines of Creation: The Coming Era of Nanotechnology*, New York, Anchor Books, 1986.

Fransman M, Junne G, Roobeek A: *The Biotechnology Revolution?*, Oxford, Blackwell, 1995.

Friedberg, E.C.: *DNA Repair*, New York, WH Freeman and Company, 1985.

Fumento, Michael: *Bioevolution: How Biotechnology is Changing Our World*, San Francisco, Encounter Books, 2003.

Gaskell G: *Biotechnology-the Making of a Global Controversy*, Cambridge, Cambridge University Press, 2002.

George S. Paul: *Beyond Humanity: Cyber Evolution and Future Minds*, Roackland, Charles River Media, 1996.

Goodman, D.C.: *From Farming to Biotechnology: A Theory of Agro-industrial Development*, Oxford, Blackwell, 1987.

Goodsell, David S.: *Bionanotechnology: Lessons From Nature*, Hoboken, Wiley-Liss, 2004.

Kurzweil, Ray: *The Age of Spiritual Machines*, New York, Penguin Books, 1999.

Moravec, Hans: *Mind Children: The Future of Robot and Human Intelligence*, Cambridge, Harvard University Press, 1988.

Morholt, E. : *Sourcebook for the Biological Sciences*, San Diego, Harcourt Brace JOvanovich, 1986.

Murray, David: *Seeds of Concern: The Genetic Manipulation of Plants*, Sydney, University of New South Wales, 2003.

Old, R.W. : *Principles of Gene Manipulation*, London, Blackwell Scientific Publications, 1989.

Postman, Neil: *Technopoly: The Surrender of Culture to Technology*, New York, Vintage Books, 1992.

Primrose, S.B.: *Principles of Gene Manipulation*, London, Blackwell Scientific Publications, 1989.

Retzer, W.J.: *Biotechnology Workbook*, Englewood Cliffs, Prentice Hall, 1991.

Rifkin, Jeremy: *The Biotech Century*, New York, Penguin Putnam, 1998

Shetty, Kalidas: *Food Biotechnology*, New York, Dekker/CRC Press, 2005.

Slater, R.J.: *Experiments in Molecular Biology*, Clifton, Humana Press, 1986.

Smith, John E.: *Biotechnology*, Cambridge, Cambridge University Press, 2004.

Switzer, R.L.: *Experimental Biochemistry*, New York, W.H. Freeman and Company, 1977.

Towle, Albert: *Modern Biology*, Austin, Holt, Rinehart and Winston, 1988.

Walden, Richard: *Genetic Transformation in Plants*, England, Open University Press, 1988.

Winston, Mark L.: *Travels in the Genetically Modified Zone*, Cambridge, Harvard University, 2004.

Index

□□□